SOUND OF WORSHIP

A Handbook of Acoustics and Sound System Design for the Church

DOUGLAS R. JONES

Focal Press
Taylor & Francis Group

NEW YORK AND LONDON

First published 2011

This edition published 2015 by Focal Press
70 Blanchard Road, Suite 402, Burlington, MA 01803

and by Focal Press
2 Park Square, Milton Park, Abingdon, Oxon OX14 4RN

Focal Press is an imprint of the Taylor & Francis Group, an informa business

Notices
Practitioners and researchers must always rely on their own experience and knowledge in evaluating and using any information, methods, compounds, or experiments described herein. In using such information or methods they should be mindful of their own safety and the safety of others, including parties for whom they have a professional responsibility.

Product or corporate names may be trademarks or registered trademarks, and are used only for identification and explanation without intent to infringe.

Library of Congress Cataloging-in-Publication Data
Jones, Douglas
  Sound of worship : a handbook of acoustics and sound system design for the church / Douglas Jones.
    p. cm.
  ISBN 978-0-240-81339-4
  1. Public worship. 2. Sound—Recording and reproducing.   I. Title.
  BV15.J66 2011
  264—dc22   2010031032

British Library Cataloguing-in-Publication Data
A catalogue record for this book is available from the British Library.

ISBN 13: 978-0-240-81339-4 (pbk)

Dedication

This work is dedicated to the memory of my father, the Reverend Lyman H. Jones. Thanks Dad, for teaching me to love God and the Church and to value hard work. Although my work was very different from yours, thank you for your support and for your pride in my accomplishments.

CONTENTS

Acknowledgments

I wish to acknowledge Columbia College Chicago for granting a sabbatical, which was spent in the preparation of this manuscript.

Special thanks to Doreen Bartoni, Dean of the School of Media Arts, Columbia College Chicago, for her support of this project.

My thanks also to my family, my wife Joy and sons Seth and Nathanael, for their patience and support during this process.

Thanks to my friend Stephen Titra for his art and photos, but most importantly for his friendship and unshakable conviction that I could do this.

Finally, thanks to Peter Gilmour DMin for his honest critique and constructive suggestions.

THE SOUND OF WORSHIP

The Christian church in the twenty-first century is a most complex entity. Indeed we may very well ask how a relatively small group of followers of Jesus Christ evolved over 2000 years into a "church" of so much diversity and schism that it is hard to recognize any kind of cohesive structure. According to Wikipedia there are in excess of 38,000 distinct denominations of the Christian church.[1] There are at least two reasons to be interested in understanding the Christian church. It is arguably the most successful religion in known history. The Church is also big business. It is a primary consumer of the services and products that acoustical firms and media systems sales and integration companies have to offer. Understanding the Church in all its complexity is critical if we are to succeed in providing the Church with services and products appropriate to its mission.

There certainly are many ways to analyze and categorize the Church. We could group churches based on a common theology, a shared history, or a shared governance philosophy. And of course any attempt at doing so would generate a considerable list of *outliers*, churches that defy categorization. It is also difficult because it is not at all clear, even to people who describe themselves as Christians, which groups are a part of the Church and which do not make the cut.

If we zoom out far enough, we can begin to see groupings, especially with respect to how church people interact with church buildings. If we further refine the study, and consider expectations for the way the rooms sound (i.e., the acoustics of the spaces and the way media, especially sound systems, are used), we can begin to see distinct patterns in the chaos.

Sound of Worship. DOI: 10.1016/B978-0-240-81339-4.00001-6

As it turns out, it is the *style* of worship that is the most accurate predictor of the kind of acoustic space and sound system required or preferred for any given church. Of course the style of worship is rooted in ecclesiology and to some degree in theology as well. There are four worship styles that are utilized by the vast majority of twenty-first–century Christian churches in America.

The first could be known by many names, but perhaps the best is the *Celebratory style*. These are the churches that follow a liturgy rich with tradition that culminates in the celebration of what some call the Mass, others the Divine Liturgy. Second, we will consider the *Evangelical style*, which focuses on the proclamation of the Word and personal salvation. The third is the *Experiential style*, which emphasizes the experience of the power of God to change lives. Finally there is the *Community style*, which is an expression of unity and commitment among believers.

As soon as we create a taxonomy of churches, someone will discover a church that does not seem to fit. This is likely true of this taxonomy as well. There are probably no churches that fit these descriptions exactly, and many churches that may fit into more than one category. This being said, the majority of American Christian churches will fit *better* in one of these categories than in the others. The architect, acoustician, and systems provider will better understand the Church as a client if these four worship styles are understood.

So how do you start with Jesus and wind up 2000 years later with this "church"? To begin to understand where all this complexity comes from and why it matters, we need to look back to the origins of the Church.

End Note

1. http://en.wikipedia.org/wiki/List_of_Christian_denominations, accessed 01/10.

2

THE EARLY ROOTS OF THE CHURCH: FROM AD 32 TO AD 313

CHAPTER OUTLINE

Academics often trace the origins of the church back to ancient Jewish and possibly even pagan practices. However, to the devout Christian, the history of the Christian church begins with a miraculous, supernatural event, the resurrection of Jesus. Most Christians mark the actual start of the Christian church with the event known as *Pentecost*, when God sent the Holy Spirit. Yet if we are to understand the development of the Church, we have to come to grips with the centrality of the resurrection. In the Christian church, Easter, not Christmas, is the most sacred day. Easter is the dynamo that powers the development of the Church and continues to energize it 2000 years after the event. It is the resurrection that makes the Christian religion different from others. The historical veracity of the resurrection is a difficult concept for modern secular minds to grasp, yet to the vast majority of Christians, the bodily resurrection of Jesus is at the center of their faith. The writer of the book of Corinthians also understood that it was difficult to accept yet critical to the faith when he wrote "if Christ be not raised, we are among all men most miserable."[1] It is the resurrection that gave the first Christians hope.

The history of the Church is not primarily a history of architecture but rather the history of a people. For Christians, the word

Sound of Worship. DOI: 10.1016/B978-0-240-81339-4.00002-8

"church" has a very important dual meaning. Church refers primarily to the people, to the congregation, the community of believers. Only in its secondary sense does church refer to a building.

ORIGINS OF THE WORD "CHURCH"

The English word "church" with its cognate *kirk* comes from the Greek word *kyriakon*, usually translated, "belonging to the Lord." *Kyriakon* can mean anything dedicated to the Lord, animate or inanimate. However, *ekklesia* is the Greek word found throughout the New Testament that is translated "church." *Ekklesia* was used in extra biblical literature to denote an assembly or congregation of free citizens summoned or "called out" (the literal meaning of *ekklesia*) by a herald in conjunction with public affairs. Even in the book of the Acts of the Apostles, Chapter 19 verse 39, *ekklesia* is translated, "lawful assembly." Nowhere is *ekklesia* translated to signify a building of any sort.

The Christians

The first followers of Jesus were the 12 men that Jesus chose to mentor or "disciple." These 12, known as the Disciples and later as the Apostles or "sent ones," were very active in spreading the teachings of Jesus known as the good news or the Gospel. Little is known about the very early Church; however from glimpses that we have in the New Testament we can put together a mosaic of sorts of what the early Church must have been like. We know that growth was rapid. In Acts 4:4 there is mention of the number of followers of Christ reaching 5000. Later we read that the number was increasing daily.[2] We know that it was largely communal in nature, that many groups held their possessions in common for the benefit of the group.[3] There was a strong sense of social responsibility, and caring for those in need.[4] We know that there were leaders, but in the early days there was not a sense of a priesthood or a segregation into clergy and laity. In fact, the Apostle Peter in his first epistle, which was a general letter to all Christians, wrote, "you are a chosen generation a royal priesthood," indicating that *all* believers had access to God, not just an elite priesthood.[5] This stood in sharp contrast to the Graeco-Roman temples where there were clear distinctions between the priests who had an inside track to the deity and the common folk who came seeking some benefit from the deity. This statement also placed St. Peter at odds with the very organized Jewish religion of the day with its stratification into many levels of devotion, spirituality, and leadership. It is also ironic that a significant part of the Christian church views Peter as the first supreme leader or Pope. The early Church was much more of a grass-roots movement. It was made up of people who at first had direct experience of Jesus or were no more than one or two degrees of separation away from Jesus.

In the early days of the Church, Christians were by necessity underground. Rome viewed the followers of Jesus as threats to the empire and the threat was met with aggressive persecution, forcing the small but dedicated group of believers to meet together for mutual support. The belief that this same Jesus who was raised from the dead would return and rescue them made it possible to face the worst that Rome could offer. The dual forces of hope and persecution were responsible for the survival of the early Christians and also for the diaspora that ultimately brought this new faith to the entire known world.

In the New Testament there are numerous metaphors that are used to describe the Church. One that is frequently used by modern Christians depicts the Church as the Body of Christ.

> *The use of this metaphor lays emphasis on the unity of the Church, the interdependence of its members and their vital relationship with its head, the Lord Jesus Christ Himself.*[6]

Early Church Architecture

In the first century AD there was no Christian architecture per se. Christian meetings were held in homes, or possibly in the burial places of the martyrs called the catacombs, although a number of historians have pointed out that the catacombs were generally far too small for any kind of meeting. The religious structures of the day were either Greek/Roman temples or they were Jewish temples or synagogues.

Three architectural elements were present in most if not all non-Christian temples of the first century. First, there was a place where the physical manifestation of the god was displayed and venerated. This was a sacred place or sanctuary in the truest sense of the word. The spiritual existence of the god was symbolized by a statue.[7] Second, there was ritual space that was open to the priests for their practices of communion with the venerated one. Third, there was a public space open to all who came to engage in some religious activity, but only through the mediation of an elite and initiated group, the priesthood.[8]

The architecture of the Jewish temple, especially the ancient temple of Jerusalem, shows marked similarity to the pagan temple with some important differences. There was a Holy of Holies or most sacred space, but there was no physical manifestation of God. It was specifically forbidden in the Ten Commandments to make any graven image of God.[9] Furthermore it was understood that God is a spirit and "no one may see God and live."[10] The presence of God was never limited to the temple; however, it could never be shut up in the sanctuary. God could be encountered anywhere. He was present in a community of laymen who both heard his words in the testimony of the Prophets and prayed to Him.[11]

By the first century AD there was the Temple in Jerusalem, where sacrifices could be made by the priesthood, but the synagogue emerged as the most common place of congregation for Jews. In contrast to the formal Temple, which contained space clearly designated as holy, the synagogue did not contain a sacred place. The synagogue contained a few symbols of God's presence but it was much more a utilitarian rather than symbolic structure. It was a space where common folk could gather around and hear the words of the Holy Scriptures spoken.

The architecture used in the Graeco-Roman cults and the Jewish religion of the day seem to foreshadow what was to evolve, but it would be many years before the fledgling Christian movement would develop an architecture of its own.

In the Christian community, there is a wide divergence of opinion on the importance of the building that houses the mystical church. Some Christians see the church (building) as imbued with a holy or sacred aspect. At the other end of the continuum there are those who see the *congregation* as having a sacred calling but see the building as just a building, referring to it not as church but as meeting house. Between these two extremes are many intermediate positions. These differences are not arbitrary. They arise out of developments over 2000 years of history and are ignored at an architect's peril.

The Church Organized

The first evidence we see of an organized Church was the Church at Jerusalem. Founded in around AD 30, it lasted until the siege of Jerusalem in AD 70 by Titus. The first Church Council was held at the church of Jerusalem in AD 49/50. There is a partial account of the Council of Jerusalem recorded in the book of Acts.[12] One of the themes that appears throughout the book of Acts and into the letters of the Apostles is the struggle with Jesus' mandate to take the Good News, or Gospel, to the non-Jews as well as to the Jews. Jesus' teaching was clearly inclusive. However, the prevalent feeling of the day was that the Gentiles or non-Jews were not worthy of salvation. The most important decision to come out of the Council of Jerusalem was that Gentiles did not have to be circumcised in order to become believers in Jesus. This is significant because it means that Gentiles did not have to become Jews in order to follow Jesus. The church in Jerusalem flourished even under persecution until Jerusalem was destroyed in AD 70 and the members had to flee across the Jordan.

Although Jerusalem may have been the foremost Christian church, it was by no means the only one. The Apostle Paul was one of the most aggressive and effective ambassadors or missionaries for the new church. Through his work there were churches founded in many cities throughout the known world of the time. By the end of his life (in AD 68/69), the Apostle Paul had visited Asia Minor, Greece, Macedonia, Crete, Rome, and probably Spain.[13] St. Paul's efforts, along with the efforts of each of the other Apostles, coupled with the fact that Jerusalem was a center of travel, ensured that the message of Jesus had spread throughout the Roman Empire by the end of the first century.[14]

The Time of Persecution

When we leave the era covered by the writings in the New Testament, we move to much thornier ground. Most of what we know about the development of the Church was written after great divisions rocked the foundations of the Church, leaving lasting wounds and disagreements. It is virtually impossible to read Church history without encountering some form of bias that makes it difficult to ascertain what really happened. We are not presenting here a scholarly work of Church history, nor are we trying to be complete. Rather, as we mentioned in the introduction, we are trying to present the high points of Church history, which we believe have bearing on architecture and therefore the acoustics and use of sound in modern church buildings. Someone has said that the history of Christianity from the time of the fall of Jerusalem in AD 70 and a century later is like a plunge into a tunnel. We know what went into the tunnel and we know what came out a century or so later—a fully articulated institution with a leadership of bishops and clergy, rituals, sacraments, and a fledgling theology. We don't really know what went on in the tunnel—at least not as much as we think we know.[15]

To set the stage for an examination of the first three centuries of the Christian church, consider the political scene of the Roman Empire. Emperor Octavian Augustus reigned during the life of Christ. According to historian Thomas Bokenkotter,

*The successors of Augustus were a strange lot. Tiberius
(d. 37), Gaius (d. 41), Claudius (d. 54) and Nero (d. 68) were
Emperors whose personal lives were darkened by bizarre,
macabre incidents and crimes: The atmosphere of their
courts was heavy with intrigue and foul suspicion. Tiberius,
under whom Christ was crucified, was a competent soldier
but an unhappy Emperor—crushed by the discovery that
his son Drusus had been murdered by his own most trusted
adviser Sejanus. His nerves shattered, Tiberius retired to
Capri where he spent the last years of his reign. Caligula was
a mentally deranged megalomaniac who was assassinated.
Claudius, weak in body and will, was dominated by his wife
Agrippa who finally poisoned him to make room for her son
Nero. Then Nero in turn murdered her and began a reign
of terror that took the lives of many of Rome's outstanding
leaders...*[16]

If there is a theme to the first 300 years or so of the Church, it
is persecution. Tertullian (ca. 140–220) was a great writer of the
early Church. He rightly observed that "the blood of the martyrs
is the seed of the Church."[17]

MARTYR

The word martyr comes from the Greek word *marturion*,
which means testimony. These were people who chose to
face violent and unbelievably cruel deaths rather then give
up or renounce their faith. Death was their testimony or wit-
ness to the power of faith. Countless thousands paid the ultimate price for their faith anonymously and although their
names were lost, the impact that their lives and deaths had
on the Church can never be forgotten.

Virtually all the leaders of the Christian movement were mar-
tyred or killed for their faith. Some names come down to us today.
We often see these names on churches or other religious buildings.

In some parts of the modern church, these early martyrs are still revered and their stories are told. It is important to remember that not all Christians were thrown to the lions. In fact, Rome was actually quite tolerant of other religions, often integrating the gods of a newly conquered nation into its own roster of gods. Rome was reasonably tolerant of Judaism as the Jews of the day kept to themselves and were not at all active proselytizers. Rome even made an exception for the Jews so they would not have to worship the Emperor as a living god.[18] At first Rome thought of Christianity as just a minor sect of Judaism. However the Christians proved to be far more trouble than the Jews ever were. They were active proselytizers and were openly defiant of the test of patriotism to Rome, namely the sacrifice to the Emperor. So they were persecuted.

Persecution of the Church began in earnest with the Emperor Nero. Most believe that he gave the order to torch Rome in AD 64; however he allegedly blamed the Christians for the disaster.

> *Informers, bribed for the purpose, declared that the Christians had set Rome on fire. Their doctrine of the nothingness of earthly joys in comparison with the delights of immortal souls in heaven was an enduring reproof to the dissolute emperor. There began a fierce persecution throughout the empire, and through robbery and confiscation the Christians were forced to pay in great part for the building of the new Rome. In this persecution Saints Peter and Paul were martyred in Rome in AD 67.*[19]

St. Ignatius.

St. Ignatius was the bishop or leader of the Church at Antioch, the chief city of Syria. Today the modern Turkish city of Antakya sits over part of ancient Antioch. Ignatius became Bishop or leader in the year 69 AD. It is in the city of Antioch that followers of Jesus first were called Christians.[20] Roughly 40 years later, the Roman Emperor Trajan came to Antioch. By some accounts Trajan was considered a good emperor, but he was no friend to the Christians. The Rev. J. C Robertson, M.A., the late Canon of Canterbury, tells the story like this:

> *When Trajan came to Antioch, St. Ignatius was carried before him. The emperor asked what evil spirit possessed him, so that he not only broke the laws by refusing to serve the gods of Rome, but persuaded others to do the same. Ignatius answered, that he was a servant of Christ; that by His help he defeated the malice of evil spirits; and that he bore his God and Savior within his heart. After some more questions and answers, the emperor ordered that he should be carried in chains to Rome, and there should be devoured by wild beasts.*
>
> *It was a long and toilsome journey, over land and sea, from Antioch to Rome, and an old man, such as Ignatius, was ill able to bear it, especially as winter was coming on. He was to be chained, too, and the soldiers who had the charge of him behaved very rudely and cruelly to him. And no doubt the emperor thought that, by sending so venerable a bishop in this way to suffer so fearful and so disgraceful a death he should terrify other Christians into forsaking their faith. But instead of this, the courage and the patience with which St. Ignatius bore his sufferings gave the Christians fresh spirit to endure whatever might come on them.*
>
> *He reached Rome just in time for some games which were to take place a little before Christmas; for the Romans were cruel enough to amuse themselves with setting wild beasts to tear and devour men, in vast places called amphitheaters, at their public games.*
>
> *St. Ignatius was torn to death by wild beasts, so that only a few of his larger bones were left, which the Christians took up and conveyed to his own city of Antioch.*[21]

Following Trajan was Hadrian (AD 117–138). Hadrian was only marginally better than Trajan from the Christians' point of view. It wasn't until the next Emperor, Antonius (AD 138–161), came to power that the Christians had any relief at all. Antonius Pius was a kindly old man when he came to power and he ordered that Christians should no longer be punished solely for their religious views.

During the reign of Antonius Pius, Justin, a Greek philosopher, became a convert to Christian thought. Instead of teaching Greek philosophy he began to teach the emerging Christian doctrines. He lived in Rome where he influenced a great number of the scholars of his day. Antonius Pius died in AD 161 and was succeeded by his adopted son Marcus Auralius Antonius. From the Christian point of view, Marcus Auralius was a good man who surrounded himself with rogues. And, although he was by all accounts a good emperor, the Christians suffered more under him than they had ever done before.[22] Justin was put to death in AD 166 and is known today as Justin Martyr.

Another name from this era that lives on is St. Polycarp. Polycarp was a friend of Ignatius, and like Ignatius was an acquaintance of the Apostle John. Polycarp was the bishop of the Church in Smyrna. The modern city Izmir in Turkey stands on the site of ancient Smyrna. Polycarp was known as a wise and holy man whose wisdom and advice were sought out by leaders of the churches throughout the Roman Empire. Polycarp (at the age of 90+) was also put to death in the year AD 166, during the reign of Marcus Auralius.

Emperor Marcus died in the year AD 181, and the Church enjoyed a brief respite from persecution. The next emperor was Alexander Severus who, although not a Christian, was friendly to the Church. Severus was murdered by Maximin and persecution returned. But it was the next emperor who managed to spread persecution of the Christians throughout the entire empire. In the year AD 249, Decius became Emperor. In January, AD 250, he published an edict against Christians: Bishops were to be put to death, other persons to be punished and tortured till they recanted.[23] Emperor Decius undid all of the few laws that offered the Christians any protection and aggressively pursued them throughout the empire. Decius was killed in battle in AD 251 and the Church once more had a brief respite.

S. POLYCARPUS.

Polycarp.

During this lull in the carnage, we see some examples of strife within the Church itself. One of the issues that was very difficult for the early Church was the sensitive issue of what to do with those who recanted their faith at the prospect of torture. There were also fierce power struggles for leadership of the Church, especially in Rome. By now (approximately AD 255) Rome had become the greatest city in the whole world. There always had been a strong Church in Rome dating back to Paul and the Apostles. By now the Church had grown quite large even under the most severe persecution that the Empire could dish out.

Cyprian was the bishop of Carthage, a city on the coast of North Africa located in present day Tunisia. Cyprian had profound disagreements with a number of leaders of the Church in Rome, especially with Stephen the Pope, or Bishop, of Rome (Pope from AD 254 to 257). Cyprian argued that all bishops were of an equal order and Rome and the leadership held no special power or authority. It was only after the death of Pope Stephen that the church in Rome and the Church in Carthage were at peace.[24] The actual scope and importance of the disagreement between these two church leaders varies wildly with historians, but it was clearly a disagreement over the power wielded by the Bishop of Rome. This is a theme that reemerges on a regular basis for the next 1500 years or so. By the end of the second century, the Church had moved a considerable distance from its original form. It was no longer a grass-roots movement of people who had known Jesus personally or were no more than one generation away from a personal, physical relationship with him. It was no longer a loose organization, a sort of brotherly society, with an emphasis on helping the less fortunate and spreading the message of Jesus. By the end of the second century, even though declared illegal by the Roman Empire, the Church had become a force to be reckoned with. The Church had become a social, financial, and political entity as well as a religious one. There was a structure, leadership, and hierarchy, and all the accoutrements of power.

THE WORD CATHOLIC

The word catholic comes from a Greek word *katholikos*, which is a compound word made up of the word *kata*, meaning according to, and *holios*, meaning whole. So, it literally means "according to the whole." It has come to mean universal or general. It is first seen in the literature in the writings of St. Ignatius. When St. Ignatius was being taken to Rome for his execution he had the opportunity to write a number of letters that have survived. In one of these, the letter he wrote to the Smyrnaeans, he refers to the Catholic church meaning the universal Christian church, which shared common beliefs and practice.

Some would call this transformation a divinely ordered progress tracing the roots of the leadership of the Catholic (Universal) church back to the Apostle Peter, viewing Peter and therefore the church in Rome, which he cofounded with the Apostle Paul to have special status.[25] Others view it as a "progressive perversion of the original Christian faith, an imperceptible back sliding of evangelical piety towards primitive religion or natural paganism."[26] Still others recognize a very human tendency to organize and battle for dominance. These power struggles, which started in the first centuries after the life of Jesus, reverberate down through the centuries and are still with us today.

In AD 257, Emperor Valerian came to power. He undoubtedly recognized that this Church was flourishing in spite of the most hideously creative persecution that the Empire could mete, and that the Church wielded real power. His tactic was to order that Christians not be allowed to meet for worship and that the leadership and the congregations be separated, figuring that by removing the leadership from the laity the Church would fall apart. Persecution started afresh, and in the year AD 258, Cyprian was beheaded. Valerian was no more successful in eliminating the Christians than his predecessors had been, and he was defeated in battle with the Persians and came to a miserable end himself.

In 261 AD, Gallienus, the son of Valerian, became Emperor. Gallienus for his own reasons sent out decrees that granted the Christians another reprieve from persecution. For approximately 40 years there was no organized persecution—but it was the calm before the storm.

By the year AD 284, the Roman Empire was getting too large for one man to handle and Emperor Diocletian reorganized the Empire and added a co-emperor to help rule. In the year AD 303 a new form of persecution arose from the pagan intelligentsia led by Porphyry. He published his work, *Against Christians,* in which he ridiculed Christ as a weakling, and insulted virtually everything the Christians held sacred.[27] This work paved the way for a new and final round of highly organized persecution. The primary reason Diocletian needed to reorganize the Empire other that its sheer size was the threats posed by the hordes of barbarians at virtually every border. The Empire was under constant threat and life was hard. It was tempting to blame Christians for all the ills, since their presence was an affront to the Roman gods. This sentiment was fueled when a priest making a public sacrifice declared that the ritual was invalidated due to the presence of Christians. Some historians suggest that nearly half of the population of the royal city of Nicomedia, the site of the palace of Diocletian, were Christian. In February of AD 303 Diocletian ordered that all places of Christian worship be destroyed. Soon after he issued three more decrees, ordering first that all Christian leaders should be imprisoned, then ordering that all prisoners sacrifice to the Emperor or be executed, then finally ordering that all Christians be required to sacrifice or face torture. This policy went on for roughly 10 years, mostly in the eastern parts of the Empire, but reached all the way to Spain, France, and even Britain.

The accounts of the torture of this decade are difficult to read. They are filled with stories of some of the most inhuman acts that can be inflicted by one human upon another. However, the Church not only survived but apparently thrived in this most hostile of atmospheres.

By the beginning of the fourth century, there were Christians in every part of the Roman Empire. There were also numerous

Christian groups in countries beyond the borders of the great Empire. There were Christians in Scotland, India, Persia, and Arabia. Many of the Goths were converted by those they captured in their raids. Around the end of the third century Gregory had gone as a missionary to Armenia where he converted the king Tiridates and convinced him to declare Christianity the official religion of Armenia. Armenia therefore holds the distinction of being the first Christian kingdom.[28]

Christianity had also reached south into Ethiopia with the founding of the Abyssinian church, which is still active today.

In terms of the religion itself, by the beginning of the fourth century, in spite of the tremendous persecution, there was a fairly well-organized creed or list of essential beliefs, there was significant agreement on which ancient writings constituted the Holy Scripture or Canon, and there was organization of the Church into a structure with the Bishop of Rome at the head and other bishops in charge of other geographical areas. There was also the beginning of a liturgy or form of worship that was virtually universal.

The fate of the Church was about to change forever as Constantine, son of one of the co-emperors, Constantius Chorus, was rising to power.

End Notes

1. Corinthians 6:17.
2. Acts 5:14.
3. Ibid., Acts 2:44.
4. Acts 9:36.
5. Ibid., 1 Peter 2:9.
6. *Zondervan Encyclopedia of the Bible* (pp. 847 ff). Grand Rapids, MI: Zondervan Corp.
7. Bieler, A. *Architecture in Worship* (p. 5). Westminster Press.
8. Ibid.
9. Genesis.
10. Genesis.
11. Bieler, A. *Architecture in Worship* (Plate 1). Westminster Press.
12. Acts Chapter 15.
13. A Church History Timeline, compiled by R. Grant Jones http://mysite.verizon.net/rgjones3/History/chronindex.htm.
14. Robertson, Rev. J. C., M. A. (1904). Canon of Canterbury, *Sketches of Church History from AD 33 to the Reformation*. London.
15. Hutchinson & Garrison. (1954). *20 Centuries of Christianity, a Concise History* (p. 36). New York: Harcourt Brace and Co.

16. Bokenkotter, T. (2004). *A Concise History of the Catholic Church* (p. 25). Doubleday.
17. *Catholic Encyclopedia*, Tertulian.
18. Hutchinson & Garrison. (1954). *20 Centuries of Christianity, a Concise History* (p. 30). New York: Harcourt Brace and Co.
19. *Catholic Encyclopedia*, Nero.
20. Acts 11:26.
21. Robertson, Rev. J. C., M. A. (1904). Canon of Canterbury, *Sketches of Church History from AD 33 to the Reformation*. London.
22. Ibid.
23. *Catholic Encyclopedia*, Cyprian.
24. Ibid.
25. Bokenkotter, T. (2004). *A Concise History of the Catholic Church* (pp. 32–34). Doubleday.
26. Bieler, A. *Architecture in Worship* (p. 22). Westminster Press.
27. Bokenkotter, T. (2004). *A Concise History of the Catholic Church* (p. 39). Doubleday.
28. Robertson, Rev. J. C., M. A. (1904). Canon of Canterbury, *Sketches of Church History from AD 33 to the Reformation*. London.

3

THE FORMATIVE YEARS OF THE CHURCH: AD 313 TO AD 1054

CHAPTER OUTLINE

Constantine

Emperor Diocletian abdicated the throne in AD 305, leaving three rivals, Maxentius, Constantius Chlorus, and Galerius to vie for power. Constantius Chlorus died the next year and his son, Constantine, was hailed as his successor (Figure 3.1). Meanwhile, Maxentius had managed to secure Rome as his stronghold. In AD 312 Constantine prepared to do battle with Maxentius for control of Rome and the West. According to the account of Eusebius of Caesarea, who was an early historian and later a confidant of Constantine, the Emperor had a vision that changed the course of his life, the Roman Empire, and most importantly the Christian church. As he prepared for battle "he saw with his own eyes the trophy of a cross of light in the heavens, above the sun and an inscription, CONQUER BY THIS attached to it…. Then in his sleep, the Christ of God appeared to him with the sign which he had seen in the heavens and commanded him to make a likeness of that

Sound of Worship. DOI: 10.1016/B978-0-240-81339-4.00003-X

Figure 3.1 Bust of Constantine.

sign which he had seen in the heavens and to use it as a safeguard in all engagements with his enemies."[1]

Constantine then took the Greek letters *charis* (χ) and rho (ρ), the first two letters of the name Christ, and used them as his personal standard. He defeated Maxsentius at the Battle of Milvan Bridge, thus securing control of Rome, and emerging as co-emperor with Licinius. The year was AD 313. With the cooperation of his co-emperor, Constantine issued a decree that legalized the Christian faith promoting tolerance of all peaceful religions throughout the Empire. Although he was not baptized into the Christian faith until much later in life, the vision or dream and his subsequent victory in battle had a profound impact on him. In AD 314 he declared to the Synod of Arles,

There were things in my own nature which were devoid of righteousness and I seemed to see as heavenly power that I might have been carrying hidden in my breast. But Almighty God, who watches from the high tower of Heaven, has vouchsafed to me what I have not deserved. Verily, past number are the blessings that He in His heavenly goodness, has bestowed on me, His servant.[2]

Constantine is an individual who still incites controversy 1700 years after his death. From a purely secular point of view, Constantine was smart enough to recognize that persecution of the Christians only resulted in their growing stronger and becoming an even greater threat to the Empire. It is likely that he saw an opportunity to exploit the amazing unity among the Christian community for the good of the Empire.[3] This, combined with a moving religious or at least spiritual experience, led him to a position of not only tolerating, but actually promoting Christianity. From the point of view of the Christians of the day, this must have been like a dream come true. The organized persecution and terror of the past three centuries was finally over. However, Constantine understood that the vast majority of his subjects were not Christian. He was careful to walk a fine line, alienating neither the pagan nor the Christian.

But what kind of man was Constantine? There are few figures from history who generate such disparate answers. The historian Burckhardt portrays him as a complicated man, boundlessly ambitious, utterly unscrupulous, "essentially unreligious…" clever in his manipulation of both Christian and pagan religions, "who persecuted what was nearest him and slew first his son and nephew, then his wife, then a crowd of friends." But based on what he accomplished, Burckhardt insists that Constantine is rightfully called the Great.[4] Others, like Grant, paint a somewhat more flattering yet still mixed portrait. "…We find a man of mixed qualities. There are divergent estimates of his statesmanship: was he weak-willed or diabolically clever? At least he was… a superb general…. A masterly cold intelligent organizer, leader, and administrator…with an immense capacity for forming schemes and carrying them out…."[5] Some look at Constantine as the one who liberated Christianity from the tyranny of Rome and who helped promote and even form the religion into a force that changed the world. Others look at Constantine as the one who virtually single-handedly merged church and state into one. The church/state dominated the history of Europe for the next millennium and beyond, and we are still coping with the effects of that action 1700 years later. Whatever descriptor we use for Constantine, the fact remains that here was an individual whose actions changed the course of history.

In AD 324, Constantine overcame Licinius and became the sole ruler of the Roman Empire. Constantine rebuilt the ancient city of Byzantium and sometime between AD 326 and 330 renamed it Constantinople. This new city became the center of the Empire, rivaling Rome as the most important city of the day. The rivalry was religious as well as political. Rome was the leader of the Western church and viewed itself as preeminent over the rest of the churches in the Empire. Rome felt that it had a right to the top position of authority in the Church, tracing that authority back to the Apostle Peter, one of the cofounders of the Church in Rome. According to their reading of Matthew Chapter 18 verse 22, Jesus had given Peter authority over the Church and that authority would be passed to the Bishop of Rome in perpetuity. However, with the founding of Constantinople, there was a new center of religious power and authority, the Eastern church. Constantine spent a fortune on lavish buildings, including temples for Christian worship. We will return to the architecture later in the chapter. When reading historical accounts of the era, we may wonder if Constantine ever regretted legalizing Christianity. Much of the rest of his life seems to be centered around settling (or trying to settle) one religious dispute after another. It is as if persecution was a sort of unifying force. When the common enemy was gone, the Church began to fight itself.

By AD 325, there was such discord in the Church over points of doctrine that Constantine was concerned about the strength of the unity that he had counted on to be the cement holding the Empire together. The primary point of controversy was the question of the Deity of Christ. Arius, an elder from the church in Alexandria, had begun to teach that Jesus was created by God but was not equal with God. This was viewed by the majority of the Church leaders as heresy; however, Arius persisted in his teaching and won considerable support—so much so that Constantine convened the first Ecumenical or churchwide council, the Council of Nicea. There were representatives from every major Church center in the Empire, over 280 bishops in all. It resulted in what is known as the Nicean Creed, a creed that is the root of the Apostles, Creed used by Catholics, Orthodox, and Protestant Christians alike.

We believe in one God the father Almighty, maker of all things visible and invisible;

And in one Lord Jesus Christ, the Son of God, begotten of the Father, only-begotten that is, of the substance of the Father, God of God, Light of Light, very God of very God, begotten not made, of one substance with the Father, through whom all things were made, both in heaven and things on earth; who for us men and for our salvation came

*down and was made flesh, and became man, suffered and
rose on the third day, ascended into the heavens, is coming
again to judge the living and the dead.*

And in the Holy Spirit.

*But those that say "There was when He was not," and,
"Before he was begotten He was not," and that "He came into
being from what-is-not," or those that allege that the Son
of God is "of another substance or essence," or "created," or
"changeable" or "alterable," these the Catholic and Apostolic
Church anathematizes.*[6]

Constantine hoped that the Council of Nicea would resolve
once and for all the problem of the Arian heresy. Unfortunately,
the problem just would not go away. The theological arguments
that Arius made, essentially subordinating Jesus to God the father,
were used to argue for subordinating the church to the state. These
discussions were never purely religious in nature; there were
always political overtones and ramifications. This issue would surface in many different ways for the next few hundred years.

Constantine was never totally successful in brokering a unity
among the diverse Christian church. But where he may have failed
in diplomacy he succeeded conspicuously as an architect. Although
there were some examples of Christian architecture before Constantine, he is credited with the most important advancement of
church building so far. There still exist examples of a wide variety of
church designs that date back to Constantine. Many of these magnificent churches owe a number of features to pagan basilicas of
the past. Pagan basilicas often were used as market, court, meet-
ingplace all at the same time. "The interior of the churches as a
whole were spacious, dignified and designed to encourage spiritual
elevation. They accomplished two things at the same time: they
recalled the disciplined Roman past which lay behind them, and
they spoke for the noble glories of heaven. It was usual, moreover,
for a Constantinian basilica to be entered from the west so that the
rising sun poured its rays of light upon the celebrating priest as he
stood in front of the alter facing the worshippers."[7]

In his book, *Theology in Stone*,[8] Kieckhefer suggests that the architecture of Constantine was an "effort to link church with state and to exalt the harmony that binds church with empire, and both the church and empire with the universe itself."[9] Perhaps the greatest example of this was the church begun by Constantine in 327, the Golden Octagon built near the imperial palace in Antioch. It was a central octagon, surrounded by a colonnade leading to a surrounding corridor. It spoke of the majesty of the emperor in the center of the empire and thus in the center of the universe.[10] Kieckhefer sees this "perhaps the inspiration for a long succession of imperial churches...."[11]

Another of Constantine's great architectural achievements was the beginning of the construction of St. Peter's basilica in Rome. Although today not much is left of the original structure, there are sixteenth century paintings showing the magnificence of Constantine's work.

In 337, Constantine, nearing the end of his life, was baptized into the Church.

BAPTISM

Baptism is still a somewhat divisive topic among Christians, so it bears looking into. Baptism is a public act that signifies joining the Church and identifying oneself as a follower of Jesus. It was practiced in the time of the New Testament, although some feel that the ritual itself was borrowed from ancient pagans. In the New Testament, baptism was done by fully immersing the penitent into a body of water. Jesus himself was baptized by John (called John the Baptist or Baptizer) in the Jordan river. Over the years some parts of the Church have held that baptism secures one's place in heaven, that it was a necessary part of the new birth referred to by Jesus in the Gospel of St. John.[12] Others held that it is merely an external symbol of an inner commitment, but had no mystical or spiritual power. Some routinely baptize infants into the Church, others feel strongly that only adults who fully

> understand the commitment that they are making should be baptized. As we shall see in Chapter 4, baptism played an important role in splintering the Church during the reformation.

In Constantine's time, the prevailing doctrine was that baptism was supposed to include a full and absolute pardon from sin. Therefore it was not unusual for adults to delay baptism as long as possible so as not to sin after being baptized. Dying in this "sinless" or fully pardoned state virtually guaranteed entrance into paradise. Constantine died the day he was baptized and was buried in Constantinople in a mausoleum he had designed for himself, flanked by statues of the 12 apostles.

With his passing, one of the most productive eras of the Roman Empire came to an end. The Empire was divided among his three surviving sons and two of his nephews, and years of bloodshed ensued.[13]

Early Ecumenical Councils

In AD 361 there was a brief resurgence of paganism when Julian "the apostate" became Emperor and proclaimed his paganism. He was killed in battle in AD 363 and with him died the last of the truly pagan emperors.

In AD 381 Emperor Theodosius called another Ecumenical Council, the Council of Constantinople. Once again the agenda was the Arian heresy. It should be noted that although most historians (even Catholics) refer to this Council as an Ecumenical, that is, "all church" council, there was no representation from Rome. Apparently this council succeeded where the Council of 325 did not. The Council of 381 marks the end of the battle with the Arians, although as noted before, the effects of the Arian controversy lasted for years.

In AD 391 Emperor Theodosius made Christianity the official religion of the Roman Empire, not just a tolerated one, thus

Figure 3.2 Jerome.

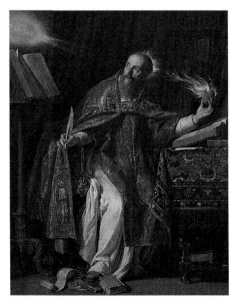

Figure 3.3 Augustine.

finishing the work Constantine started and essentially fusing church and state into one all-powerful agency.

Jerome and Augustine

Mention should be made here of two individuals who were very important in the development of the Christian Church in the fourth century. The first is Jerome (Figure 3.2). Born of wealthy Christian parents in AD 331 he was privileged to study in some of the best schools of his day. His most important contribution to the development of the Church was the translation of the Scriptures (what we now call the Old Testament) into the Latin used in his day. This translation is known as the Vulgate and is still used by biblical scholars today.

Another luminary of this period was Augustine (Figure 3.3). Born in AD 354 in what is now Algeria, North Africa, of Berber origin, he studied in a number of schools, learning Latin and Greek. He is most well known for two important works: his *Confessions*, which is sort of an autobiography, and his later work, the enormous 22 volume *City of God*, which took him 13 years to write.

Council of Chalcedon

Although the next years saw a number of minor and not so minor squabbles and disagreements, the next important milestone came 60 years after Christianity became the religion of the Empire. Of course we mustn't overlook Patrick starting his mission to Ireland in AD 423, but in AD 451, Emperor Marcian, the emperor of the east, called another all-church council. This council was to meet in the city of Chalcedon, just across the Bosporus from the imperial city, Constantinople. Marcian wanted a council that would "end disputations and settle the true faith more clearly and for all time."[14]

There were approximately 520 bishops in attendance. The deck was stacked in favor of the East as all but four of the bishops were from the eastern sections of the Empire. There were two delegates from Rome and two from North Africa.

This fourth ecumenical council was important for three reasons. First, it addressed the question of the nature of Jesus. In this aspect the Council of Chalcedon was a tremendous success as it set forth the doctrine of the Divinity of Jesus that has stood the test of time. From a strictly historical point of view however, the Council of Chalcedon marks the onset of an important split in the Church. The split was not complete, but the relationships were seriously strained. The Church in Alexandria, Egypt, was a very powerful entity at the time, but was fiercely independent of both the East (Constantinople) and the West (Rome). In the words of the Coptic (Alexandrian) Church itself,

> *The Coptic Church was misunderstood in the 5th century at the Council of Chalcedon. Perhaps the Council understood the Church correctly, but they wanted to exile the Church, to isolate it and to abolish the Egyptian, independent Pope, who maintained that Church and State should be separate. Despite all of this, the Coptic Church has remained very strict and steadfast in its faith. Whether it was a conspiracy from the Western Churches to exile the Coptic Church as a punishment for its refusal to be politically influenced, or whether Pope Dioscurus [Bishop of Alexandria] didn't quite go the extra mile to make the point that Copts are not monophysite, the Coptic Church has always felt a mandate to reconcile "semantic" differences between all Christian Churches.*[15]

The rift was serious, and although the Church in Alexandria, now known as the Coptic Church, was invited to participate in subsequent ecumenical councils, this fractured relationship is still felt by the Coptic Church today. Finally, the Council of Chalcedon was the arena for an important debate over the role of the Roman Church. For reasons that are difficult for modern readers to grasp, it was very important to the leadership of the Church in the fifth century to rank the importance of the various Church centers. All agreed that Rome should be ranked first, although it is not clear exactly what the ramifications of being *first* were. In what is known as Cannon 28, the Council of Chalcedon ranked the Churches as follows: Rome, Constantinople, Alexandria, Antioch, and Jerusalem. And even though the relationship of Alexandria with the whole was strained, the bishop of each of these regions was given the title Patriarch.[16]

Pope (Bishop, Patriarch) Leo of Rome did not accept Cannon 28. He felt that it threatened the primacy of Rome even though the canon listed Rome first. Bokenkotter, in his *Concise History of the Catholic Church*, observes that the dispute was really over two forms of primacy, "directional vs. administrative." "The directional primacy—the right to be the final court of appeals in matters of faith… certainly belonged to Rome… {but there was another type} a patriarchal or administrative type…. Constantinople had certainly won the right to such a primacy."[17] Constantinople and Rome patched things up but the stage was set for the great rift some 600 years in the future.

The year AD 476 marked the end of the western empire as Flavius Odoacer elected to rule in Rome as a lieutenant of the Emperor in Constantinople.

Rise of Monasticism

For the sake of brevity, we move ahead to around AD 525 where we find the monk Benedict. Benedict was not the first monk, but his work transformed the monastic life. Monasticism is a very important aspect of the Christian faith. According to Mark Noll in his book *Turning Points*,[18] "the rise of monasticism was the most important—and in many ways the most beneficial institutional event in the history of Christianity.... In the Centuries between Constantine and the Protestant Reformation almost everything in the church that approached the highest, noblest, and truest ideals of the gospel was done either by those who had chosen the monastic way or by those who had been inspired...by monks." The roots of Monasticism seem to be in Egypt. As early as the second century there is evidence of monastic religious orders in Alexandria.

MONASTICISM

Monasticism is a religious way of life where the participants (known as monks) usually cut themselves off from society and dedicate their lives to prayer and spiritual discipline. The Monastery of Christ in the Desert defines it this way:

Defying a simple definition, Christian monastic spirituality is primarily an approach to God in response to God's invitation found in Sacred Scripture: "Seek first the Kingdom of God" (Mt 6:33). Monastic spirituality implies a single-heart (solitary) seeking of God. This may or may not be carried out in the company of others, (the monastic tradition has embraced both), but the focus is clearly on returning to God, and making use of certain specific practices: prayer, fastings, silence, vigils, reading, good works.[19]

Benedict, most likely in response to what he felt was a corrupt monastic system even as early as AD 525, composed his *Regula*, or Rule. This Rule is meant to give order and structure to a monastery and to safeguard against all forms of corruption. Benedict's Rule may seem harsh by modern standards but it is regarded as the foundation for modern monasticism and is still at the core of every Benedictine monastery around the world. As we move into the Middle Ages, we will come back to Monasticism as its impact on the development and spread of the Church should not be underestimated.

In the year AD 532, Emperor Justinian began construction on the Great Church of Constantinople. Known as the Hagia Sophia or Holy Wisdom, the cathedral took five years to complete. The church measures roughly 253 feet by 259 feet with a gigantic dome that sits some 203 feet above the floor. The dome has a diameter of close to 108 feet. On December 27, AD 537 the church was dedicated by Patriarch Menas, and in AD 553 was the site of the fifth Ecumenical Council.[20] Some of the artwork dating back to the sixth century is still on display in the church (Figure 3.4).

In AD 570, Mohammed was born in Mecca, a commercial and religious center in Arabia. In AD 610 he received his first revelation; an angel commanded him to "recite" and he began teaching what had been revealed to him about Allah. Soon a new religion was born called Islam. In AD 622 he and his followers

Figure 3.4 The Hagia Sophia in Modern Times.

were pushed out of Mecca by a rival tribe. They fled (known as the *Hegira*, the beginning of the Muslim calendar), but as is so often the case, persecution and exile backfired. In Medinah, Mohammed, the prophet of Allah, made many converts. By the time of his death in AD 632 he had converted roughly a third of Arabia.[21] In two years virtually all of Arabia had converted and by AD 642, when Islam moved into Egypt, virtually all of Arabia, Persia, Syria, Palestine, and some of India were Muslim. This also marked the beginnings of intense enmity between Islam and Christianity. It gradually escalated to a peak just after the turn of the millennium.

ISLAM

Islam and Christianity have a number of things in common. Many of the Old Testament stories also appear in the Holy Book of Islam, the Qur'an. In fact, both the Jews and the Arabs trace their ancestry back to Abraham. Isaac, Abraham's son with Sarah, is the father of the Jews. Ishmael, the son of Abraham and Hagar, is the father of the Arab people. Of course there is a distinction between the Arab people and the religion Islam. There are Arabs who are not Moslem and of course Moslems who are not Arab. The conflict that exists between these two great religions is not likely to be resolved any time soon. Islam and Christianity both teach exclusive paths to God. They both are aggressively evangelistic and they both claim that the other is a distortion or an incomplete manifestation of the truth.

This is not a work of comparative religions and we are not going to attempt to analyze the complex relationship or history that Islam and Christianity share. The spread of Islam did play an important role in the history of the Christian Church and therefore must be mentioned here.

The Church in Northern Europe

The fifth and sixth centuries marked the movement of Christianity into northern parts of Europe. In AD 601 Augustine (the monk, not the theologian of AD 354) became the first archbishop of Canterbury. The Church in England was firmly integrated with the church in Rome at the conference of Whitby in AD 664. By now Gaul (present day France) was mostly Christian. The general area including modern Germany, however, was still pagan. In the AD 720s two English missionaries were commissioned by the Bishop of Rome (by now generally known as the Pope) to bring Christianity to Germany. Willibroad went to Utrecht in the Netherlands and Winfrid (also known as Boniface) went on to Germany. Both were rather successful not only in making converts but also in setting up church structures and connecting them firmly to Rome.[22] This was an era of expansion for the Church; it was also a time when the Church, especially the Church in Rome, was evolving into a super power in its own right. There is evidence that the term Pope had been used as early as the second century, referring to the Bishop of Alexandria. By the AD 600s the term was used more and more as a title reserved for the Bishop in Rome. One of the great Popes of this era was Gregory I (AD 590–604). Gregory was a consummate theologian and reformer of the Western Church. For the purposes of this book, it should be noted that Gregory I had a tremendous impact on worship in the Church, especially regarding the use of music. He promoted the use of plain song chant, which was a beautiful blend of communication, art, and architecture. Since most of the laity were illiterate, chant provided them with a way to hear and remember the Scripture. It was also a way to integrate the artform of music into worship. From an acoustical perspective, the Gregorian chant is a fascinating study. Many have postulated that the chant evolved as a way to communicate in the highly reverberant cathedrals of the day. By singing in the dominant modes or resonances of the building one essentially harnesses the room itself to aid in communication. There is no evidence to suggest that this was a conscious effort on the part of Gregory I or any of the other early proponents of chant. Whether by design or serendipity it remains to this day as an example of symbiosis between music and acoustics that is rarely excelled.

In AD 787 the last of the truly ecumenical councils was held in Nicea. Known as the Second Council of Nicea, this council addressed the use of icons or painted images in churches. For years the Church struggled with the role of art in the Church. Because of the prohibition against graven images that appears in the Ten Commandments (Exodus 20:4) and because the use of images and such played a key role in pagan worship, Christianity has always been nervous about including "sacred images" in church buildings. The second Council of Nicea addressed the issue by noting that there was a clear distinction between *honoring* an image used to call to mind an event or Biblical teaching, and the *worship* of such an image. This was very important to the Eastern Church, which developed a very elaborate and beautiful collection of icons, many surviving to this day. Nicea II is also significant as this was the last time that the East and the West met in a spirit of unity.

The rise of Charlemagne to power is a fascinating read. He was a Frank (from modern France) who wound up being crowned Emperor of Rome by the Pope. As Noll says, this raises three important questions: (1) How did the Pope come to have the power enough to crown a Roman Emperor? (2) How had the king of the Franks risen to a position to be so crowned? (3) And how did this new relationship between the Pope and the greatest ruler of northern Europe shape the centuries-long period of western history usually referred to simply as Christendom?[23] We will not attempt to answer these questions here. Clearly the relationship between church and state had come a long way from the days of Emperor Diocletian, who we saw briefly in the last chapter. We might wonder what Diocletion would have thought if he were told that 500 years in the future, his successor to the throne would be crowned by the leader of the very church he tried so desperately to abolish. From a very practical point of view, this new relationship of church and state raised the question of authority. Where did the final authority lie, with the Church or the state? As we shall see later, this was a question that would plague the history of Europe for the next thousand years.

By the end of the first millennium, The Holy Roman Empire was in full force; although showing signs of decay, Christianity had reached into all parts of Europe including Russia. Church and state were one, but tensions were mounting in the Church. The East and West, Constantinople and Rome, were growing further apart on a variety of issues, some religious, some political. The Universal or Catholic Church was about to go through its first major split.

End Notes

1. Quoted in Noll, M. (1997). *Turning Points: Decisive Moments in the History of Christianity* (p. 50). Baker Books.
2. Quoted in Grant, M. *Constantine the Great* (p. 141). New York: Macmillan Publishing.
3. Grant, M., *Constantine the Great* (p. 51). New York: Macmillan Publishing.
4. Quoted in Hutchinson, P. *20 Centuries of Christianity* (p. 49). New York: Harcourt Brace & Co.
5. Grant, M. *Constantine the Great* (p. 105). New York: Macmillan Publishing.
6. Quoted in Hutchinson, P. *20 Centuries of Christianity* (p. 57). New York: Harcourt Brace & Co.
7. Grant, M., *Constantine the Great* (p. 193). New York: Macmillan Publishing.
8. Kieckhefer, R. (2004). *Theology in Stone*, Oxford Press.
9. Ibid., p. 42.
10. Ibid., p. 42.
11. Ibid., p. 42.
12. John.
13. Grant, M. *Constantine the Great* (p. 226). New York: Macmillan Publishing.
14. Quoted in Frend, W. H. C. (1984). *The Rise of Christianity* (p. 770). Philadelphia: Fortress Press.
15. Taken from the St. Takla Coptic Church website, /st-takla.org/Coptic-church
16. Ware, Timothy. (1997). *The Orthodox Church* (New Edition, p. 26). Penguin Books.
17. Bokenkotter, T. (2004). *A Concise History of the Catholic Church* (p. 92). Doubleday Press.
18. Noll, M. (1997). *Turning Points: Decisive Moments in the History of Christianity* (p. 84). Baker Books.
19. From www.christdesert.org/, Christ in the Desert Monastery website, accessed 6/09.
20. From www.patriarchate.org/ecumenical_patriarchate/hagia_sophia, accessed 6/09.
21. Noll, M. (1997). *Turning Points: Decisive Moments in the History of Christianity* (p. 118). Baker Books.
22. Hutchinson, P. *20 Centuries of Christianity* (p. 91), New York: Harcourt Brace & Co.
23. Noll, M. (1997). *Turning Points: Decisive Moments in the History of Christianity* (p. 110). Baker Books.

CELEBRATORY WORSHIP

4

THE CELEBRATORY WORSHIP STYLE
Worship the Lord in the Beauty of Holiness

A sacrament is a festive action in which Christians assemble to celebrate their lived experience and to call to heart their common story. The action is a symbol of God's care for us in Christ. Enacting the symbol brings us closer to one another in the Church and to the Lord who is there for us.[1]

Those Christians who belong to the Orthodox, Catholic, and Episcopal churches all practice similar forms of Celebratory worship. It is not the intent here to fully parse the differences that exist between these three great Churches, although the history of the divisions that occurred, ultimately defining each one, will be presented in later chapters. Each of these churches sets forth well-considered arguments defending their own particular form of Celebratory worship.

Sound of Worship. DOI: 10.1016/B978-0-240-81339-4.00004-1

The Catholics refer to their celebration as the Mass. There is some discussion as to the ultimate source of this word but most Catholics trace it back to a Latin word *missio*, which is the root of the English word mission. The word means to be sent out or dismissed. In early times there was a point during the church service when the uninitiated were dismissed. Gradually, popular speech used the name of one of the parts of the ritual, the dismissal or the missio to indicate the whole.[2]

The Orthodox do not use the term Mass. Although to the outside observer there seems to be much in common with the Roman Catholic Mass, it is inappropriate to use the Latin name for the Eastern worship service. In the Orthodox Church, the celebration is referred to as the Divine Liturgy. The word liturgy is from a Greek word meaning common action or service, and its use signifies that this is an act of common worship that unites the faithful in service to God.

The Anglican use the term Mass to describe the worship service, but it should be very clear that although again to the outsider there is very little discernible difference between an Anglican Mass and a Roman Catholic Mass, to the faithful in these two churches there are very important differences that must be respected.

The Celebratory form of worship is by far the most ancient of all the forms of worship. At the heart of the celebratory worship is the observation of the Eucharist. The Eucharist is the most Holy and certainly the most important part of the service. It is the point in the service when the last supper is commemorated by the sharing of bread and wine. It is this focus on the Celebration of the Sacrament of Eucharist that sets apart this form of worship. It sits in sharp contrast to the other forms of worship as we shall see.

The Sacraments

Most categorizations of the Christian church divide the Church broadly into two camps, the Sacramental and everybody else. Indeed this style of worship could have been named Sacramental. The word Celebratory was chosen as it connotes more of a style of worship rather than Sacramental, which is really a theological perspective that makes the Celebratory style possible and gives it life. At the risk of oversimplification, a Sacramental understanding of God's relationship to humankind is one that believes that "God delights to use tangible, concrete earthy means—matter itself—to communicate His grace, redemption and presence to us...."[3] When a believer partakes of or participates in a Sacrament, something real, if mystical, happens. The Christian is changed, grace is imparted. When we undergo the sacrament of baptism for example, it is much more than a symbolic exercise. The baptized is welcomed by God into the community of the faithful. It is this Sacramental world view, if you will, that determines virtually everything about the Celebratory church, from the architecture of the buildings to the way the services are conducted.

There are seven sacraments recognized by the Roman Catholic Church, and different numbers of sacraments recognized by the other Sacramental churches. Even if they disagree over the number of Sacraments they do agree that the Sacrament of the Eucharist is central to the life of the Church and the spiritual well being of the Christian. To the Christian who worships in the Celebratory style, participation in the Eucharist is a mystical experience of partaking of the body and blood of Jesus, who in the Last Supper distributed bread and wine and told his disciples to take and eat, this is my body broken for you, this is my blood shed for you. Partaking in the Sacrament causes something to happen. There is a spiritual nourishment, a feast for the soul. Participation in the Eucharist is an act of faith and transcends the rational mind. In fact, in the Orthodox church the Sacrament is often referred to as the Mystery and the Eucharist is sometimes called the Mystery of Mysteries.

At this point it is important to reflect on this. Readers will undoubtedly react to the notion of a sacrament and the Eucharist differently depending on their religious perspective. Christopher Hall states in his book *Worshiping with the Church Fathers*:

> *It is sad and ironic that the sacrament of the Eucharist*
> *continues to divide Christians. Interpretations of Christ's*

invitation to communion with himself in bread and wine and through him with other members of his body have often led to misunderstanding and suspicion, conflict, hatred, violence and death among Christ's own disciples. Too often the Eucharist has led to schism rather than unity. Sadly, disagreements over the meaning of the Eucharist continue to divide Christians to this present day. Surely the Lord is not pleased with this state of affairs.[4]

For those who were brought up Catholic, Orthodox, or Episcopal (especially for those who continue to participate in the life of the Church), this sacramental world view is norma-

St. Michael's Church, Chicago, Illinois.

tive, and life-giving. For those who were brought up in a nonsacramental part of the Christian church, essentially everybody besides the Catholic, Orthodox, and Episcopal, this sounds foreign and is contrary to everything you have been taught about Christianity. For those who are totally secular, or those who were raised in a religion other than what is considered broadly Christian, this may sound like some sort of bizarre cannibalism. It is not the intent of this book to convince anyone of the validity of one form of worship or one theological position over another. Rather we explore the different styles in order to better understand how to better serve the Church in all of its diversity. It is critical for architects, acoustical engineers/consultants, and sound system designers/installers to understand that the differences that exist between the different worship styles are profound and will affect the way that your services are utilized. This understanding of the

Sacraments not only defines the style of worship, but also defines the architecture and therefore the acoustics and use of media.

Sacred Space

In all the Sacramental or Celebratory churches, the Mass or Divine Liturgy is to be conducted in consecrated or sacred space. As we shall see, this position stands in sharp contrast to the other forms of worship. What makes a space sacred? On one level, a space is made sacred by an act performed by someone whose office entitles them to sanctify a space for holy use. In this sense any space may be consecrated and made holy. Most Christians would agree that God transcends any building that we might construct for worship. And yet is seems as though there are attributes of certain structures that seem to speak to us in special ways that suggest holy or consecrated space. The theme of sacred space will be revisited in Chapter 7.

In his book about church architecture, *Theology in Stone*, Richard Kieckhefer[5] offers four factors when considering the design of sacred space: Spatial Dynamics, Centering Focus, Aesthetic Impact, and Symbolic Resonance. Although Kieckhefer broadly applies this analysis to all churches, even to those who do not view the physical church building as having any sort of sacred element, it is useful to look at the introductions to his first four chapters as a way to begin to understand what sacred space means.

Spatial Dynamics: "Entering a church is a metaphor for entering into a spiritual process: one of procession and

St. Michael's Church, Chicago.

return, or of proclamation and response, or of gathering in community and returning to the world outside. The form of sacred architecture will follow largely from the conception of spiritual process it is meant to suggest and foster, the type of dynamism it aims to promote."[6]

Centering Focus: "Entering a church is a metaphor for centering one's attention and one's life on some particular purpose, usually represented by some object vested with real and symbolic importance in and beyond the act of worship...."[7]

Aesthetic Impact: "Entering a church is a metaphor for entering into the presence of the holy.... Writers on church architecture are in broad agreement that a church's aesthetic character is central to its purpose: that beauty is 'the only symbol we humans can devise that illuminates the transcendent'; that the climate of liturgy is one of awe, mystery, wonder, reverence, thanksgiving and praise, requiring nothing less than the beautiful...."[8]

Symbolic Resonance: "Entering into a church is a metaphor for entering into a shared world of symbolic narratives and meanings, somewhat like entering into a story and discovering the richness and internal coherence of its structure. The symbolic associations of a church's structure, furnishings and decoration evoke a sense of the sacred, as its aesthetic qualities elicit a sense of the holy."[9]

In the Celebratory style of worship, symbolism is very important. Virtually all aspects of the architecture and furnishings have meaning and support the Mass. It is the only one of the worship styles in which the visual is an important part of the act of worship. Indeed in some forms of the Celebratory worship, all five senses are involved. All the symbolism of the altar, vestments, furnishings, indeed of the architecture itself, appeal to the eye. The sounds of the bells, chants, prayers, spoken and musical responses are a central part of all Celebratory worship. In the Eucharist the worshippers partake of the sacred bread and wine. In the passing of the peace, worshippers embrace or shake hands with each other.

Suggested Listening

Listen to the sacred music recording at www.sound-of-worship. com.

St. Joseph Orthodox Church, Wheaton, IL.

The Worship Service

It is important for anyone who is intending to serve the Church as architect, acoustical engineer, or sound systems designer to realize that the three forms of the Celebratory church do not agree with each other in ways that are very important to each of the three. Care must be taken to not offend by making assumptions because on the surface they look very similar. Although there are significant theological, ecclesiological, and historical differences, from a strictly acoustical and audio perspective they are similar enough so that what will be appropriate for a Roman Catholic church will likely be applicable to Orthodox and Episcopal churches as well. Intending no offense to the Orthodox or Episcopal churches, we will be using the Roman Catholic Mass as an example of how Celebratory worship is conducted. In most Celebratory churches the Liturgy will change somewhat depending on the church calendar. What follows is a sort of generic Mass for illustrative purposes only.

The Mass has four parts consisting of an Introductory Rite, Liturgy of the Word, Liturgy of the Eucharist, and Concluding Rite.

Introductory Rite

The Mass opens with a procession into the church. The priest and attending ministers enter the church and process to the alter generally accompanied by music. Sometimes incense is used. The ministers are seated and the priest then addresses the gathered congregation with words of greeting, "In the name of the Father and of the Son and of the Holy Spirit." The congregation replies, "Amen," making the sign of the cross. From this point until the end of the Mass, the worshippers are considered to be praying to God. During the Introductory Rite there will be a blessing upon those assembled, as well as prayers of confession of sin, and recitations that may be spoken, chanted, or sung.

Liturgy of the Word

There are three distinct readings from scripture every Sunday Mass. Two of the readings will be from parts of the Bible other than the four books known as the Gospels, Matthew, Mark, Luke, and John. These reading are announced, read, and then the people respond, "Thanks be to God." After each reading is a response often in the form of a psalm, which is sung by the cantor and the people. The two readings are followed by the presentation of the Gospel. On special Sundays and especially in those groups who observe more elaborate forms of the Mass, this is a point of high pageantry where a special copy of the Gospel is processed into the church accompanied by incense and music. All will stand for the reading of the Gospel. After the Gospel is read and the people make the appropriate response, the people will sit and the priest will present the homily of the day. The priest will usually focus his remarks on the scriptures read during the Mass, but he may speak on any relevant topic. After the homily, the people once again stand and there is a recitation of a Profession of Faith, which will be either the Nicene Creed or the Apostles' Creed. After the profession, the priest will begin a time of general intercession or prayer. Each petition or request is followed by the people's response, "Lord hear our prayer."

Liturgy of the Eucharist

This is the high point of the Mass. The priest will prepare the bread and the wine, consecrating each in turn, with prayers of blessing. The prayers and responses may be either spoken or sung. The Eucharistic Prayer is then recited, which is essentially a dialog between the priest and the people. Following the Eucharistic Prayer, the people will recite the Lord's Prayer and then pass the "sign of peace" to each other. This may be an embrace or handshake or some culturally appropriate gesture of welcome and peace. The passing of the peace is followed by the breaking of the bread and the preparation of the communion vessels. When all is ready, the faithful are invited to come to the front and receive the bread and wine from the ministers. It is expected that only those who are "in communion" with the Church, that is, those who are members in good standing, are invited to partake. In some traditions, mostly Orthodox, before the Eucharist begins, those who are not going to be participating in the Eucharist are invited to leave. While the people are coming forward to receive the bread and wine, there will often be music played or sung by the choir. After all have been served, there will be a moment of silent reflection, after which there is a prayer, which ends the time of the Eucharist.

Concluding Rite

After the Eucharist, there is a blessing and prayer for the people, and the dismissal.

Those Christians who celebrate the Mass or Divine Liturgy every Sunday often find deep meaning in the symbolism of the rituals. If you belong to one of these groups, you can go anywhere in the world and attend a Mass or Divine Liturgy and feel right at home. There was a time, of course, when Catholic, meaning universal, really meant Catholic and all Christian churches were alike, but to find that we have go back quite a few years.

End Notes

1. Gus, T. (1981). *The Book of Sacramental Basics* (p. 53). New York: Paulist Press.
2. Catholic Encyclopedia, accessed 2/2010.
3. Hall, C. A. (2009). *Worshipping with the Church Fathers* (p. 21). Downers Grove, IL: InterVarsity Press.
4. Ibid., p. 51.
5. Kieckheffer, R. (2004). *Theology in Stone.* Oxford University Press.
6. Ibid., p. 21.
7. Ibid., p. 63.
8. Ibid., pp. 97,98.
9. Ibid., p. 135.

5

THE CATHOLIC–ORTHODOX SPLIT

It is naive to suggest that there was little change in the Christian church in the first thousand years. Yet compared with the diversity we see today, the first millennium ended with a remarkably unified entity. At no other time in the history of the Church has the term catholic (meaning universal) been more aptly used. The persecutions, debates, and power struggles had not managed to destroy the unity that was so powerful and attractive to men like Constantine. This was all about to change forever.

The Great Schism

On July 16 in the year 1054, as the Saturday afternoon Mass was about to begin in the Hagia Sophia, the great cathedral in Constantinople, Cardinal Humbert of the Roman church along with two other emissaries from Pope Leo IX of Rome walked into the church, disrupting the Mass. They placed on the altar a document that excommunicated or expelled from the Church Michael Cerularius, who was the Patriarch of Constantinople, the Emperor Michael Constantine, and all of their followers. Humbert and companions then turned and left, ceremoniously shaking off the dust from their feet. The document itself was inflammatory with many unfounded accusations and the mode of delivery

Sound of Worship. DOI: 10.1016/B978-0-240-81339-4.00005-3

unpardonable. This act caused an uproar in Constantinople, which was quelled only by the public burning of the document. A few days later, Cerularius responded by declaring Humbert anathema, thereby expelling him from the Church, but as noted by Timothy Ware (Orthodox historian and Bishop) in his history of the Orthodox church, the west did not include the entire Roman church in its statement.[1] Of course Humbert did not act out of the blue. Pope Leo IX had led a military expedition, defending the Papal states in southern Italy against the Normans. This exercise was a failure and Pope Leo IX was captured and held prisoner for a number of months.

Patriarch Cerularius, an outspoken critic of the West, was angered that the Pope would interfere in areas in southern Italy that were pro East. In response, Cerularius closed down the few Latin churches that remained in Constantinople.[2] In addition he fanned the flames of an ancient controversy dividing East from West, namely the use of unleavened bread for the Eucharist. The West held that the bread used in the Eucharist must be unleavened—containing no yeast. The East held that the use of leavened bread was good and right. Cerularius asked a colleague to write a letter blasting the Latin use of unleavened bread and that letter found its way to the Pope. The letter written in Greek was translated for the Pope by Humbert, whose limited knowledge of Greek probably made it sound even more contentious than it was. Humbert, then, was dispatched to Constantinople in response to Cerularius' recent activities.[3] By all accounts this tragic confrontation might have been avoided if two less quick-tempered and hot-headed men had been involved. Thus, 1054 is cited as the date of the Great Schism between East and West. The events of July 16, 1054 notwithstanding, the schism was clearly many centuries in the making. The year 1054 did not start the rift nor did it mark the point of irreconcilable differences. Mark Noll traces the origins of the schism back to the second century, noting the differences between two great Church leaders of that time, Clement of Alexandria (ca. 150–ca. 215) and Tertullian of Carthage (ca. 160–ca. 225).

These differences are worth another look, since the divergent mentalities of the two notable leaders, although separated by only a few hundred miles on the North African coast, would loom larger and larger as the centuries passed. Again, Tertullian's first language was Latin, Clement's was Greek. Tertullian boldly challenged the pagan culture of his day with the realities of the Christian faith, while Clement sympathetically sought aid for Christianity from the best that paganism offered. Tertullian coined new words (like "Trinity") and was eager to construct formulas of the faith (regula fidei) *which he then expected to end theological debate. Clement meditated at great length on the truths of the faith and used formulas to stimulate discussion about the ultimate realities of Christianity. Tertullian was a lawyer, Clement a philosopher. Tertulian reasoned toward action, Clement reasoned toward truth.*[4]

This profound difference between a Latin and Greek approach to Christianity was at the center of many of the disputes that we have seen thus far, and certainly was present in this great schism. In the West, the Pope had power and authority that superseded the government. This was not true in the East where the Emperor still wielded more power than the Church. The West had achieved unity in the Church whereas the East seemed comfortable with the notion of divided authority. From the East's point of view the Bishop of Rome was one of five patriarchs who had essentially equal footing.[5]

Ware writes:

East and West were becoming strangers to one another, and this was something from which they were both likely to suffer. In the early Church there had been unity in the faith, but a diversity of theological schools. From the start, Greeks and Latins had each approached the Christian Mystery in their own way. At the risk of oversimplification, it can be stated that the Latin approach was more practical, the Greek more speculative; Latin thought was influenced by juridical ideas, by concepts of Roman law, while the Greeks understood theology in the context of worship and in the light of the Holy Liturgy. When thinking about the Trinity, Latins started with the unity of the Godhead, Greeks with the three-ness of the persons.... Like the schools of Antioch and Alexandria, within the East, these two distinctive approaches were not in themselves contradictory; each served to supplement the other and each had its place in the fullness of the Catholic tradition. But now the two were becoming strangers to one another—with no political and little cultural unity, with no common language—there was a danger that each side would follow its own approach in isolation and push it to extremes, forgetting the value in the other point of view.[6]

So the rift occurred. Historians from both traditions are quick to point out that at first, the break was at the top and the laity knew very little about it. There were attempts made to heal the rift, but none were successful. Roman Catholic historian Thomas Bokenkotter writes:

...since Pope Leo died before the legates' arrival in Constantinople, the legates had been acting without actual authority, so it would have been easy for a subsequent Pope to repudiate their action without any loss of prestige. The whole episode could have been glossed over as a mere unfortunate lapse. And at first it was viewed this way in the East. But in the West it was different. With Humbert's star rising in the Curia, his version of the affair—which sounded like a hymn of triumph—was accepted as the right one. Subsequent Popes followed his line. Pope Gregory VII was Humbert's closest friend and would not have dreamed of repudiating his action. Humbert, they believed, did the right thing excommunicating an unrepentant and contumacious bishop; since the Patriarch's successors also refused to seek absolution, they too were viewed as partaking in the schism...the whole patriarchate of Constantinople was eventually included in the excommunication.[7]

This mutual excommunication remained in effect in spite of two efforts to heal the break, in 1274 and again in 1439. In 1963 the Second Vatican Council (Vatican II) finally took steps towards reconciliation.[8] The rift remains, however, with little movement to bring together the congregations of East and West. Today there is the Roman Catholic church, which is conspicuously headed by the Pope, and is the Western or Latin church of old. The Eastern tradition includes the ancient patriarchates of Constantinople, Alexandria, Antioch, and Jerusalem, and more recent churches of Russia, Romania, Serbia, Bulgaria, and Georgia. The Orthodox churches of Cyprus, Greece, the Czech Republic, Slovakia, Poland, and Albania are also part of this group. Each of these Orthodox churches is governed by a Patriarch—sometimes known as a Metropolitan—that has Pope-like authority over its own. They are in a loose relationship with each other, recognizing the Patriarch of Constantinople as a "first among Equals."[9] As recently as in the year 2000 more attempts were made to bring these two great bodies closer together, but they are still far apart.

To make matters more complicated there are also churches known as the Eastern Catholic churches. These are *Eastern* churches who maintain a full relationship with the Roman Catholic church and should not be confused with Eastern Orthodox. The use of the word Orthodox in the name of a church usually denotes it as one that is not in a relationship with the Roman Catholic church.

To the secular observer who compares a Roman Catholic Mass to an Orthodox Divine Liturgy there are a few differences, but not as many as we might expect. One important difference is that in many Orthodox churches, the congregants stand for the entire service! There are some real doctrinal differences but to the nonreligious person these seem hardly worth the fuss. If we look closely at the Creeds, for example, we will find three words (one word in the original Latin) that appear in the Western (Catholic) but do not appear in the Eastern Creed. The addition of these words (the "Filioque" controversy) by the West did as much to divide the two as anything else. To these two great churches that together represent most of the Christians on the planet, the details do matter!

Many authors agree that it was the Crusades and not the events of AD 1054 that actually cast the rift in a form that lives on today. So to better understand the historical forces that keep Catholic and Orthodox apart, we briefly turn to one of the darkest chapters in the history of Christianity, the Crusades.

The Crusades

No attempt will be made here to present the Crusades with all the complex social, economic, political, and religious dynamics that gave rise to them. Citing selected outcomes of some of the Crusades should not be construed to be ignoring or trivializing the tremendous impact that they all had on both European history and the history of the Middle East, or on certain groups, especially the Jewish community.

According to the Church History Timeline, in 1009 the Moslem leader Fatimid Caliph Hakim ordered the destruction of the Church of the Holy Sepulcher. By 1014, Hakim had burned or pillaged 30,000 churches.[10] So by 1095, when Pope Urban II announced the first Crusade at the council of Clermont in France, the Moslem threat was real and was felt throughout the Christian church. It is interesting to read the words of Urban II as he called for action:

> *Although, O sons of God, you have promised more firmly*
> *than ever to keep the peace among yourselves and to preserve*
> *the rights of the church, there remains still an important*
> *work for you to do. Freshly quickened by the divine*
> *correction, you must apply the strength of your righteousness*
> *to another matter which concerns you as well as God. For*
> *your brethren who live in the east are in urgent need of your*
> *help, and you must hasten to give them the aid which has*
> *often been promised them. For, as the most of you have heard,*
> *the Turks and Arabs have attacked them and have conquered*
> *the territory of Romania [the Greek empire] as far west as*

the shore of the Mediterranean and the Hellespont, which is called the Arm of St. George. They have occupied more and more of the lands of those Christians, and have overcome them in seven battles. They have killed and captured many, and have destroyed the churches and devastated the empire. If you permit them to continue thus for awhile with impunity, the faithful of God will be much more widely attacked by them. On this account I, or rather the Lord, beseech you as Christ's heralds to publish this everywhere and to persuade all people of whatever rank, foot-soldiers and knights, poor and rich, to carry aid promptly to those Christians and to destroy that vile race from the lands of our friends. I say this to those who are present, it is meant also for those who are absent. Moreover, Christ commands it.

Over a thousand years ago, the leaders were using the "domino theory" to motivate the public! But just in case this call to arms by Pope Urban II was not enough he went on to add the next section:

All who die by the way, whether by land or by sea, or in battle against the pagans, shall have immediate remission of sins. This I grant them through the power of God with which I am invested. O what a disgrace if such a despised and base race, which worships demons, should conquer a people which has the faith of omnipotent God and is made glorious with the name of Christ! With what reproaches will the Lord overwhelm us if you do not aid those who, with us, profess the Christian religion! Let those who have been accustomed unjustly to wage private warfare against the faithful now go against the infidels and end with victory this war which should have been begun long ago. Let those who for a long time, have been robbers, now become knights. Let those who have been fighting against their brothers and relatives now fight in a proper way against the barbarians. Let those who have been serving as mercenaries for small pay now obtain the eternal reward. Let those who have been wearing themselves out in both body and soul now work for a double honor. Behold! on this side will be the sorrowful and poor, on that, the rich; on this side, the enemies of the Lord, on that, his friends. Let those who go not put off the journey, but rent their lands and collect money for their expenses; and as soon as winter is over and spring comes, let him eagerly set out on the way with God as their guide.[11]

This was powerful stuff to eleventh century ears. The promise of immediate remission of sins for all who died in battle was not to be taken lightly. The common man of the day lived in mortal fear of divine retribution for his sins. As the Crusade got underway a number of would-be leaders viewed Urban's call to arms as a license to do whatever necessary to achieve the goal, including raiding villages and taking whatever was needed for the actual crusade. It was during these preparations that many of the atrocities occurred. In 1096 a band of Crusaders making their way from France south ultimately to Jerusalem, passed through the German town of Worms. There they killed some 800 Jews, many of whom had sought refuge with a local bishop to no avail.

The Crusaders captured Antioch in AD 1098, then went on successfully "liberating" town after town until they achieved the ultimate prize, Jerusalem, in AD 1099. In a sense it is unfortunate that the first Crusade was so successful. It was followed by at least seven other crusades over the next few centuries, most of which were far from successful by any metric.

According to Ware, it was the fourth Crusade that brought the division between the two churches to the level of the people. In the year 1204 the Crusaders were bound for Egypt. They were persuaded by the son of a dispossessed Emperor to make a small detour and aid in returning him to the throne at Constantinople. The Crusaders agreed and proceeded to sac the holy city of Constantinople. To add insult to injury the Crusaders then set up a Latin ruler. The Church in the East has never fully forgotten those three days of pillage.[12] The schism of 1054 was political but 1204 made it personal. For the Greeks it was no longer about doctrinal differences argued over by leaders in ivory towers. The West had acted with aggression and sacrilege, and the Greeks hated them for it. Constantinople remained under Latin rule until 1261 when the Greeks reclaimed it.

Most historians count eight crusades of this period. The tragedy is that after all that bloodshed, effort, and resources the Crusades ultimately succeeded only in sowing seeds of ill will that live on to today. None of the great holy cities that were "freed" by the Crusades remained in Christian hands for very long. The crusades left no one unscathed. Christians (of East and West), Moslems, and Jews all were deeply wounded by these events and those wounds still have not healed all these years later. Kenneth Latourette, the Protestant historian from Yale says in his book, *Christianity Through the Ages*, "The net effect of the Crusades was to weaken the Byzantine Empire, to deepen the gulf between the Eastern and Western wings of the Catholic church and permanently to embitter relations between Christians and Moslems. The Crusades are a striking example of that *perversion of the devotion* [emphasis mine] evoked by Christianity which has been seen again and again."[13]

On April 7, 1453, the Turks began to attack Constantinople. The defenders were outnumbered 20 to 1 but managed to hold off the invaders for seven weeks! "In the early hours of May 29th, the last Christian service was held in the great Church of the Holy Wisdom. It was a united service of Orthodox and Roman Catholics, for at this moment of crisis...(they) forgot their differences. The Emperor went out after receiving communion and died fighting on the walls. Later that same day the city fell to the Turks and the most glorious church in Christendom became a mosque."[14]

End Notes

1. Ware, T. (1977). *The Orthodox Church* (p. 59). Penguin Books.
2. McBrien, R. (1997). *The Lives of the Popes* (p. 175). San Francisco: Harper.
3. Bokenkotter, T. (2004). *A Concise History of the Catholic Church* (p. 137). Doubleday.
4. Noll, M. (1997). *Turning Points* (p. 135). Intervarsity Press.

5. Hutchinson, P., & Garrison, W. (1959). *20 Centuries of Christianity* (p. 106). New York: Harcourt Brace and Co.
6. Ware, T. (1977). *The Orthodox Church* (pp. 48–49). Penguin Books.
7. Bokenkotter, T. (2004). *A Concise History of the Catholic Church* (pp. 138, 139). Doubleday.
8. Ibid., 141.
9. http://www.religioustolerance.org/orthodox.htm
10. http://www.geocities.com/Heartland/Pines/7224/Rick/chronindex.htm
11. http://www.fordham.edu/halsall/source/urban2-fulcher.html
12. Ware, T. (1977). *The Orthodox Church* (pp. 60–61). Penguin Books.
13. Latourette, K. S. (1965). *Christianity through the Ages* (p. 136). Harper Chapel Books.
14. Ware, T. (1977). *The Orthodox Church* (p. 61). Penguin Books.

6

CATHOLIC REFORM

In the last chapter we saw the Church was no longer Catholic, but divided into the Eastern and Western segments. Skipping ahead some 400 years, we find ourselves at a very low point in the history of the Roman or Western church. The Protestant Reformation is covered in Chapter 9, and we shall see that it was an attempt to repair or maybe salvage what was left of a profoundly corrupt fifteenth century Catholic Church. However, it is wrong to think that the Catholic Church remained in the sad state that it found itself in during Luther's time. To fully appreciate the modern Catholic Church we need to look at two final historical events; the council of Trent and Vatican II.

Response to the Reformation

As the movement that Luther started gained momentum, the Catholic Church realized that the world had changed forever. They had lost the absolute sway they held over a large part of Europe. Although they would never admit it openly, and they would continue to condemn his positions and practice, they knew Luther was right at least insofar as it was clear that something had to change. In 1545 Pope Paul III convened the Council of Trent. It met over a 15-year period in three major sessions.

Sound of Worship. DOI: 10.1016/B978-0-240-81339-4.00006-5

Trent was a certainly a response to the Protestant movement and it was the most visible part of what some refer to as the Catholic "counter-reformation." Other writers see it as simply a Catholic Reformation whose time had finally come. Still others have used the term's reaction and revival to describe this period in the history of the Catholic Church.[1] Certainly all of these terms apply, and the appellation of this time in the history of the Church is still debated by historians. Whatever it is called, by the end of the sixteenth century, the Catholic Church had changed dramatically.

Like most events in history, the Council of Trent cannot be fully understood in isolation. In fact, there were attempts at reforming the Church that predate the Council of Trent by a few decades. Remember that the purpose of this very concise history of the Church is to examine those events that caused schism in the Church and are largely responsible for the diverse state of the modern Christian Church. We have ignored two great centers of European Catholicism—Spain and Italy—largely because they were not the epicenters of any major break in the Church. However, in order to get a better appreciation for the Council of Trent and other attempts at reform, we briefly need to look at the Church in Spain and Italy.

In Spain in the late 1400s, under the rule of Ferdinand and Isabella, there were a number of reforms and controls of the excesses of the Catholic Church, which in a medieval sort of way were responsible for keeping the Spanish Catholic Church strong. Nowhere in Europe were the state and Church more inexorably entwined then in medieval Spain.[2] This of course was a heady time for Spain. Columbus had "discovered" a new world under the Spanish sponsorship and this and other conquests had lined the coffers of the state, making Spain a force to be reckoned with. The state had successfully put down external threats from Islam, and the Church had successfully suppressed virtually all internal heresy via the Inquisition. According to the historian Williston Walker,

> ...Spain was witnessing vigorous reformatory work led by
> Queen Isabella and Cardinal Ximenes. It combined zeal for a
> more moral and intelligent clergy, abolition of glaring abuses

and Biblical studies of the learned not *[emphasis mine] for the people, with unswerving orthodoxy, judged by medieval standards, and repression of heresy by the Inquisition. It was this movement that was to give life and vigor to the Roman revival often, though rather incorrectly, called the Counter Reformation. Outside of Spain it had very little influence when Luther began his work. Indeed the decline of the Roman Church was nowhere more evident than in the feebleness with which it met the Protestant onslaughts during the first quarter century of the Reformation, and the incapacity of the Popes themselves to realize the gravity of the situation and to put their interests as heads of the Church above their concerns as petty Italian princes.*[3]

Segments of the Church in Italy were eager for reform as well, but not for a revolt. A number of religious orders were formed during this time with the express purpose of trying to repair the Church from within. In fact, in 1517 while Martin Luther was nailing his theses to the door in Wittenburg, a religious order called the Oratory of Divine Love was founded in Rome, dedicated to the cause of reform.[4] Among the leaders of this group was Gian Pietro Caraffa, who had spent a number of years in Spain and admired the efficiency and efficacy of the Spanish Church in dealing with matters of reform. Largely through his efforts, the Inquisition was implemented in Italy as well and any hope for a Protestant movement was eliminated. Caraffa was later to become Pope Paul IV and continued his pursuit of reform while Pope.

There were actually a few attempts at a reconciliation between the Catholic Church and the unorganized Protestant movement. Perhaps the most significant occurred in 1541. Pope Paul III convened a colloquium at Regensburg, a town in southern Germany. This meeting was attended by moderate Catholic leaders who wished to heal the rift in the Church and a number of Protestant leaders. On the agenda were a number of talking points, mainly of doctrine. The group actually managed to agree on two of the points, but it blew apart not surprisingly over two points, transubstantiation, and the Catholic assertion that only the leadership of the Church had the right to interpret scripture.[5] Regensburg was viewed as a failure and this was, in some measure, the reason for convening the Council of Trent a few years later. It was a time for fixing those things that were broken but more importantly restating, so there could be no doubt, those elements of Catholic dogma that were immutable.

The makeup of the Council was a point of great controversy. It was certainly not by accident that the majority of the delegates to the Council of Trent were from Spain and Italy. These were the ecclesiastical centers that either had not experienced the threat from the Protestants or had acted swiftly to squelch any Protestant movement.

The Council of Trent was certainly not an attempt to reconcile with the Protestants but rather it was a much needed time for Catholic introspection. Nonetheless, Protestant leaders were invited to attend some of the proceedings.

There were numerous problems facing the Catholic Church that were on the table during the council of Trent—matters of doctrine, fiscal abuses, and administrative failures to name a few. The Council addressed these problems with vigor. In a series of decrees they responded to Luther by clearly defining what is Universal or Catholic and what is not. "Scripture and Tradition were *both* [emphasis mine] declared necessary in determining the faith of the Church."[6] Equally important was a statement that only the Church had the authority to interpret the scripture, effectively disallowing the translation of the scripture into the vernacular. There were clear statements on the sacraments: all seven were deemed to be indispensable for salvation. These seven sacraments were to be administered only by clergy.[7] This was in direct response to the Protestant position which held that clergy were not necessary, as according to the Protestant interpretation of scripture (especially 1 Peter 2:9[9]), all believers are "priests." The Council also put forth statements regarding the hierarchy of the Church, the priesthood and more. One of the most important outcomes was the appointing of a commission to reform the Mass.

THE SEVEN SACRAMENTS

The word sacrament comes from the Latin word *sacramentum* meaning "a sign of the sacred." In Catholic tradition and practice there are seven sacraments. These are actually seven ceremonies in which Catholics believe Christians experience the saving presence of God. They are considered to be both signs and instruments of God's grace. The seven sacraments are Baptism, Eucharist, Reconciliation, Confirmation, Marriage, Holy Orders, and Anointing of the Sick. They are—with rare exception—performed only by a Priest.

Baptism: Performed generally on infants by sprinkling holy or consecrated water on their heads. It is held that Baptism sets the recipient free from original sin.

Eucharist: Also known as Communion, it is the commemoration of the Last supper where Jesus passed around bread and wine and told his disciples to eat and drink in memory of Him (Luke 22: 7–20). In the Catholic tradition the bread (an unleavened wafer) and consecrated wine are distributed to those who are confirmed in the Church. The Catholic Church holds that Jesus is actually present in the elements (transubstantiation).

Reconciliation: The ceremony by which one repents of sin and receives God's forgiveness through the aegis of the Church. Before Vatican II this was known as Confession.

Confirmation: A commitment made by mature individuals (not infants or very young children); it is a sort of initiation into the full life of the Church.

Marriage: The union of a man and woman in a lifelong commitment.

Holy Orders: Also known as ordination, the process whereby a priest takes vows, which essentially initiates him into the priesthood.

Anointing of the Sick: Formerly called Last Rites, a ritual of healing for spiritual, mental, and physical illnesses.[8]

The Roman Catholic historian Bokenkotter writes:

The commission that Trent set up to reform the Mass did their work rather quickly, and in 1570 issued the Missale Romanum, *which was made binding on the universal Church and which remained virtually unchanged until the 1960s. Its introduction marked a new era in the history of the Mass: in place of the allegorical Mass there would now be the rubrical Mass—the priest being obligated under penalty of mortal sin to adhere to its most minute prescriptions. Here again it is the extreme conservatism of the council that strikes the eye. It is at least conceivable that they might have taken a creative approach. They might, for instance, have introduced the vernacular as the Protestants had done so successfully. But instead they acted defensively and protectively. One reason for this is that in the polemical climate of the times they could not afford to admit that the Protestants could be right about anything. This would impugn the claim of the Roman Church to divine authority.*[10]

The few decades of Catholic reform did not result in the sort of changes that Protestants would have wanted. They did however result in a Catholic Church with a clear position and a stand against the excesses and corruption that had sparked the Protestant revolt. The work that the Council of Trent did stood for over 400 years and was not revisited with any real vigor until Vatican II, which was convened in the 1960s.

Vatican II

In 1962, the Roman Catholic Church under Pope John XXIII called for the second Vatican Council. The first Vatican Council adjourned in

1870 leaving much of its charge undone. With the second Vatican Council, known as Vatican II, Pope John XXIII wished the Council "to increase the fervor and energy of Catholics, to serve the needs of Christian people." Once, when asked to more fully explain his intentions for Vatican II, Pope John got up and opened a window, letting the fresh air blow in.

Vatican II tackled some of the most difficult questions facing the Catholic Church, and it was not without controversy. Conservative Catholics wanted desperately to preserve the heritage of the Catholic tradition with all that implies, and protect it from modernism, whereas liberals hoped it would indeed be a breath of fresh air to an institution that was all too stagnant. In four separate sessions spanning three years, the Council issued 16 formal documents on topics ranging from doctrinal positions to practical concerns. It addressed marriage, birth control, and the role of laity in Church life. It was also quite ecumenical in its attempt to heal old wounds. It acknowledged that there are true Christians who are not part of the Catholic Church. It also took important steps toward healing the rift with the Eastern Orthodox Church. On a practical level, things like the prohibition against eating meat on Friday was lifted. The Mass itself also went through a rather dramatic change. This last topic gets much more attention than the rest of the decisions as it touched all Catholics where they lived. Before Vatican II the Mass was said in Latin, the universal language of the Church. Now it was permissible and even encouraged to celebrate the Mass in the language of the people. Part of the restructuring of the Mass was an increased emphasis on part of the Mass that involved the priest talking directly to the congregation. The sermon, renamed the homily, now played a more important role in the Mass.

In addition, it addressed the use and type of music permissible in the Church, and although the pipe organ was recognized as the preeminent sacred instrument for worship, it allowed others to be included.[11] Suffice it to say, before Vatican II the notion of a polka Mass now popular in parts of Wisconsin, would not have been possible! It was this council that opened the door for the use of sound reinforcement in Catholic churches and indeed made it necessary. The story is told of a consultant working in a large Catholic church, taking acoustical measurements for the design of a new sound system. He was approached by an elderly man who asked what he was doing. He replied that he was designing a new sound system for the church. He looked at the consultant wistfully for a moment then replied in his rich Irish brogue, "Ach laddie, I dunna need to hear the prayers. I've been hearing the prayers since I was twelve. But I suppose it would be nice once to hear the homily!"

For all the changes that Vatican II did initiate, it is almost silent on the subject of the architecture of the Roman Catholic church. The only direct mention of the church structure says that "…when churches are built, let great care be taken that they be suitable for the celebration of liturgical services and for the active participation of the faithful."[12] The phrase "active participation of the faithful" has been interpreted to mean all sorts of things, and triggered a wave of renovations to traditional churches and new churches being built. There has been considerable reaction to some of the modern architecture. Writers like Steven Schloeder refer to modern Catholic architecture as banal and insists that churches should not be built simply as places for people to gather, but for the worship of God.[13] Others like Michael Rose were even more outspoken in their criticism of modern church architecture. The title of his book published in 2001 says it all: *Ugly as Sin: Why They Changed our Churches from Sacred Places to Meeting Spaces and How We Can Change Them Back Again.*[14]

As can be expected, the effects of Vatican II are still being assessed and debated. Some were very slow in accepting all the changes permitted by the Council. There are still Catholic churches who hold Latin Mass. For others it did not go far enough. We are probably still too close to it to fully appreciate the full impact that Vatican II had on the Church and on society. Mark Noll writes:

Because of its very size and the weight of its traditions, what happens in the Catholic Church profoundly affects the direction of Christian history in general. When, in addition, leaders like John Paul II appear, whose preparation for the papacy included the ravages of war, the rigors of life under Communism and intensive training as a philosopher, the prominence of the Catholic Church becomes even greater. How the Second Vatican Council comes to be judged, how its adjustments to Catholic tradition work out in the Church's future course, may one day be viewed as a critical turning point in the twentieth-century history of Christianity.

End Notes

1. Grimm, H. J. (1973), *The Reformation Era 1500–1650* (p. 301). NY: Macmillan Publishing Co.
2. Walker, W. *A History of the Christian Church* (3rd ed., p. 286). NY: Charles Scribner's Sons.
3. Ibid., p. 374.
4. Grimm, H. J. (1973). *The Reformation Era 1500–1650* (p. 304). NY: Macmillan Publishing Co.
5. Noll, M. A. (1997). *Turning Points: Decisive Moments in the History of Christianity* (p. 206). Baker Books.
6. Bokenkotter, T. (2004). *A Concise History of the Catholic Church* (p. 242). Doubleday.
7. Grimm, H. J. (1973). *The Reformation Era 1500–1650* (p. 326). NY: Macmillan Publishing Co.
8. http://www.americancatholic.org/features/sacraments/default.asp, accessed 9/09.
9. "But you are a chosen people, a royal priesthood, a holy nation, a people belonging to God, that you may declare the praises of Him who called you out of darkness into His wonderful light." 1 Peter 2:9 New International Version.
10. Grimm, H. J. (1973). *The Reformation Era 1500–1650* (p. 224). NY: Macmillan Publishing Co.
11. See *Sacrosanctum Concilium 1963,* Chapter IV, Sacred Music, articles 120, 121.
12. Ibid., Chapter VII, article 124.
13. Schloeder, S. J. (1998). *Architecture in Communion: Implementing the Second Vatican Council Through Liturgy and Architecture.* San Francisco: Ignatius.
14. Rose, M. S. (2001). *Ugly as Sin: Why They Changed Our Churches from Sacred Places to Meeting Spaces and How We Can Change Them Back Again.* Manchester, NH: Sophia Institute Press.

7

THE ACOUSTICS OF CELEBRATORY CHURCHES

CHAPTER OUTLINE

Case Studies

We now turn our attention to the acoustics of the Celebratory church. In Chapter 6 we briefly covered the second Vatican Council or Vatican II. Vatican II marks a significant turning point in the Catholic church, which of course is reflected in the architecture of Catholic churches built after 1963. The other main celebratory churches, the Episcopal and the Orthodox, did not go through such sweeping reforms so there tends to be a bit more consistency in the architecture of these churches. There are more Celebratory churches built in traditional styles then there are post Vatican II Catholic churches built in more modern or experimental forms. It is also the case that the Celebratory church is not the fastest growing church in the United States. Sound system and acoustical designers will likely find more work in older Celebratory churches desiring renovation or upgrades than they will in new construction.

As we saw in Chapter 4, the whole focus of the celebratory worship service is the administration of the Sacrament of the Eucharist. Everything in the church building is subservient to this one event. We saw that in this tradition the space is sacred. The question of how to express the sacred in tangible materials is an age-old one. It is interesting to look at what the Bible has to say about sacred space. As we mentioned in Chapter 1, in the English Bible when we see the word *church* it never refers to a building. The church always refers

Sound of Worship. DOI: 10.1016/B978-0-240-81339-4.00007-7

to a people. In the Old Testament however, God did dictate how the tabernacle, the forerunner of the Jewish Temple, was to be built. This is the only biblical description of how to build a holy space. God's design for the space that he would eventually inhabit for the purpose of interacting with His people can be found in the book of the Exodus, starting with the specifications in the twenty-fifth chapter and the actual building of the Tabernacle starting in the thirty-fifth chapter. We are struck at the frequency of the use of the word "skill" in this context—over 20 times in the next few chapters and always when describing the workmanship, or the abilities of the craftsmen working on the project. It is clear that God expected this holy place to be built with great skill and craftsmanship and that it was to be built out of the finest materials available. All of Kieckhefer's four factors mentioned in Chapter 4 were present in the design of the tabernacle. The spatial dynamics in the way the tabernacle was divided into sections are of increasing "holiness."

The tabernacle was divided into four distinct areas: the outer court for the general population, the Holy space for the Priests, and the Holy of Holies reserved for the high priest only, and then only once a year. This hierarchy of space reflected the way God interacted with the people, always through priests as intermediaries. The centering focus was clearly the veil or curtain that separated the Holy from the Holy of Holies. The aesthetic impact is evident in the emphasis on the quality of the work. And finally the whole design is symbolic on many levels.

The Old Testament description of holy space seems to confirm Kieckhefer's statement that humans rely on beauty to illuminate the transcendent. However, beauty alone does not fully convey the sense of the transcendent. Scale also contributes, if not directly to the sense of the holy, at least to a sense of awe, which is a part of the whole experience. There is a reason that those things we name the wonders of the world are all large. Standing next to the Great Wall of China or the Pyramids of Egypt may not invoke a sense of the presence of the divine, but it does provoke awe, and at very least a realization that at some point in history someone felt passionate enough about something to warrant this sort of output of effort. This is certainly part of the experience of walking into one of the great cathedrals. The scale, beauty, and majesty of the building all seem to point to something transcendent, to something other. Even totally secular people who have no religious context recognize that there is something special about these buildings.

If in the Sacramental church the architecture must reflect the sacred and the holy, then the acoustics that stems from the architecture must also reflect the same. This of course is a much more difficult question. In architecture, and in most of the visual art forms, there is at least a language rich with metaphor that can be used to describe how the visual might evoke or elicit the sacred or the holy. This language is woefully lacking in the field of acoustics; indeed it is lacking in things aural in general. In English, which has more words by far than any other language, there are very few words that describe aural experience uniquely. The few that do, like *silent, quiet, cacophonous, shrill,* and *sibilant,* either describe the *absence* of sound or undesirable properties of sound. Enter into a grand cathedral and what words come to mind to describe the acoustics of the space? We are forced to borrow terms from other senses or resort to metaphor or onomatopoeia. We might be able to speak of sacred architecture but can we speak of sacred acoustics? Can we speak of sacred sound? It is beyond the scope of this book to try to unravel the reasons for this lack of vocabulary, much less to propose a language.[1] Yet if there is such a thing as sacred space then there must exist also an acoustic space, which if not sacred at least lends itself to sacred activity. Steven Webb, in his book *The Divine Voice, Christian Proclamation and the Theology of Sound,*[2] which addresses the role of sound in the modern Church, speaks of a Theo-Acoustics. He quotes Walter Ong in observing that "sound, not paper, is the native medium of communication."[3] He also observes that "the sacramental theology of Roman Catholicism instructs the faithful in the habits of an intimate and loving hearing. We are able to listen to each other only if we learn to listen to God. After all, the priest has the power to verbally transform matter and to forgive sins. The spoken word elevated our merely physical presence to each other into something deeply spiritual."[4] In the Sacramental church, sound is important in a way distinct from the other Christian worship styles. Sound is a sacred thing. It has a power that is traced back to creation itself when God spoke the world into existence.

If sound is so important, what can we do in the realm of acoustics to create or preserve an environment that is conducive to worship in this tradition? The first concrete step we must take is to insure that "we do no harm," and to realize that in the Celebratory church the architecture trumps everything! It is as important to avoid doing harm to the acoustic space as it is to take active steps to produce an acoustically appropriate space. One form of harm of course is noise. The problem of noise is covered more fully in Chapter 21. Of all the forms of worship, the Sacramental is the most likely to be disturbed by extraneous noise. The design team should strive for noise levels of no higher than NC25. This can be a challenge in an urban setting, especially if there are many large windows incorporated in the architecture.

Another form of harm is to not respect the reverberation as something desirable. Acousticians sometimes tend to view reverberation as the enemy. Certainly there are situations where the reverberation time is excessive and needs to be creatively controlled. However, that perception of scale mentioned earlier is experienced through two senses. The eye sees the large structure and is impressed. The auditory system at the same time is detecting the initial time gap, which corroborates and reinforces the sense of scale, contributing to the impression of awe. Chapter 21 will deal with the topic of reverberation more fully.

Many Celebratory churches still utilize pipe organs in their sanctuaries. Of all the types of churches in the United States, we are most likely to find a functional pipe organ in a Celebratory church. If a pipe organ is being considered for a new facility, it will likely dominate the discussion of the acoustics of the room. On some level it is reasonable to allow a million dollar plus investment to determine the acoustics of the room, but we should never lose sight of the fact that the purpose of the space is not just the organ. Pipe organ installers and "voicers" or tuners often speak the language of acoustics but sometimes the words get in the way. For example, an excerpt from a pamphlet on the acoustics required for a pipe organ suggests that there be at least 250 cubic feet of volume for every seat in the room (sanctuary) and that the reverb should be even between 63 Hz and 8000 Hz. "Even reverb" is a fairly ambiguous term. If "even" means that there would be the same reverb time at 63 Hz as there is at 8000 Hz, that is neither feasible nor desirable. Virtually all rooms large enough to house a pipe organ will have a rising reverberation time at the low frequencies. Ultimately this is not a performance space, this is holy space, which exists primarily to serve the Mass. Other uses of the space such as organ recitals are secondary uses.

Perhaps the most offensive thing we can say about a Celebratory worship service is to call it a performance. It is *not* a performance! It may look like one to an outside observer, but it is a sacred act of worship. It contains pageantry and even some choreography and music, but it is not a performance. It is not about the people who are "performing" the Celebration. It is about the Church leading its own in the Sacrament of the Eucharist. For most churches that worship in the Celebratory style, anything that hints of performance or commercialism is viewed very suspiciously. Churches may install lighting systems to light the Altar and enhance the beauty of the sanctuary, but balk at having the controls of the lighting system be anywhere in the congregation, much less have someone "running lights" during a Mass. It also explains why for the most part it is very difficult, if not impossible, to get a Celebratory church to install a mixing console in the congregational seating area and have someone "mix" the service. Celebratory services are often held in the most difficult of acoustic spaces and require sophisticated sound systems to get intelligible speech to the congregation. Furthermore most parishes offer Masses at different times during the week where the numbers of those in attendance can range from very few to a full house. The different Masses are often officiated by different ministers. All these conditions should be taken into consideration by those running the sound system, yet the systems in Celebratory churches rarely benefit from human control. The systems are most often automatic and represent significant compromise.

One of the serious challenges facing the Celebratory church in the twenty-first century is something that many leaders of the Church have not yet articulated but grapple with on a weekly basis, namely exploring new ways to worship while maintaining the traditional architecture. Many Catholic churches, and Episcopal churches as well, in an attempt to be more inclusive, community minded, and relevant, have begun celebrating Mass specifically for minority groups. For example, St. Nicholas Church in Evanston, IL, celebrates a Hispanic Mass every Sunday afternoon. The Mass attracts a population of Catholics who want to incorporate their traditional music into the Mass. They bring guitars of various types as well as a variety of percussion instruments. The problem is that St. Nicholas has a mid-band reverberation time, half full, in excess of three seconds! Such a long reverberation time makes it difficult, if not impossible, to use percussive musical forms. This is not a problem that can be solved by simply installing a better sound system. It is music that is essentially incompatible with the acoustics of the space. Or, from another point of view, it is an acoustic space incompatible with the music.

A consultant was called in to suggest solutions. The consultant who was not brought up in the Catholic or Celebratory style of worship first suggested some modest acoustical treatment. This suggestion was met with immediate veto for fear that it would ruin the sound of the pipe organ, to say nothing of the visual aesthetic of the space. The consultant then suggested that the only solution was to move the Hispanic Mass into the fellowship room. The consultant might as well have suggested erecting a statue of Martin Luther in the sanctuary! The Mass would *not* be celebrated in the fellowship room! St. Nicholas still struggles with trying to use instruments other than piano and organ in its musical repertoire. This is a problem that most Catholic churches built before Vatican II face in an ongoing way. There are no simple solutions. Still, from the point of view of the Church, the Mass was celebrated, it was effective and valid. The participants received grace through the Sacrament of the Eucharist, even if the drums sounded muddy and it was hard to understand the singers.

To summarize the acoustics and sound system needs of Celebratory churches:

In pre-Vatican II churches the predominant acoustic issue is likely to be reverberation. The reverberation may be well suited to the pipe organ and chant but ill suited for everything else. Suggest treatment very judiciously. Not only will there be push-back due to the way the treatment will look, but it is likely that congregations worshipping in pre-Vatican II churches have come to expect the church to sound a certain way. To them it most certainly is not broke, so don't fix it! Perhaps the ultimate solution for Celebratory worship would be a church with variable acoustics that could be adjusted for a mid-band reverberation time of around three seconds for pipe organ music and chant but be turned down to around a second or less for contemporary or alternative forms of the Mass. Deal with the intelligibility problems with very high Q or at least very directional loudspeaker systems that can also easily be integrated into the architecture. Steerable array speaker systems like the IntelliVox® (Figure 7.1) are very useful in the older Celebratory churches. The architecture will trump everything. If the sound system can be guaranteed to provide everyone with perfectly intelligible sound, but is deemed an eyesore, it will not be implemented. These churches need technology to be as inconspicuous as possible and in general will not tolerate anything that appears to suggest that the Mass is a performance.

If an older church is considering a renovation, or in the case of new construction, the noise goal should be a PNC25 at least, with PNC20 preferred.

If the church can afford it, reverberation enhancement systems like the LARES® system can be very effective at giving the church the best of both worlds. A church can be built or treated so that the natural reverberation is low, and the artificial

Figure 7.1 Use of IntelliVox® Column in a Large Reverberant Catholic Church.

Suggested Listening

Listen to the LARES Demo at www .sound-of-worship. com.

system can be turned on to create the sense of live-ness and reverberance that is expected. When more modern music is desired, the artificial system can then be turned down to a minimal setting to allow the music to be played.

Case Study

St. Rose of Lima Catholic Church, Miami Shores, Florida

Electro-Acoustical Consultants: The Audio Bug Inc.

St. Rose of Lima Catholic Church (Figure 7.2) was completed and dedicated in 1961, pre-Vatican II. It reflects a more modern approach to church architecture than the traditional style common in Catholic churches of its day, indicating that perhaps the Archbishop and architect were somewhat more forward-thinking than some.

The design reflects the fact that this is a South Florida structure. The Nave is designed to fill with light during the day, a common theme in Roman Catholic churches. Sunlight streaming through the stained windows (plastic, not glass, to avoid damage during hurricanes) fills the room with a wide hue of colors. At night, interior lighting creates a similar though somewhat more mysterious environment.

Figure 7.2 St. Rose of Lima Catholic Church.

Architecturally, the room was definitely designed with acoustics in mind. This is not your typical "shoebox-shaped" room. There are very few parallel surfaces and a healthy dose of diffusion is found everywhere. The side walls taper inward toward the center of the space and the plan view of the space resembles a shallow teardrop, with the walls at the entrance (rear) closer together than those at the altar end of the room. An open brick lattice wall envelopes the altar, providing a great deal of diffusion. Two short transepts extend outward from either side of the front section of the seating area just forward of the altar. One has been closed off with moveable wood and glass partitions as a "cry room" and the other remains a contiguous space to the main body of the room.

The room's broadband T_{60} is just over two seconds and supports the organ, choir, and congregational singing very well. A recent renovation, in which old carpeting was replaced with Italian tile, increased this metric somewhat but did not prove to be a liability.

Speech intelligibility is very good; analysis of RIR (Room Impulse Response) measurements indicate STI (Speech Transmission Index) of 0.53 (%Alcons = 9.4334). The sound system consists of four Tannoy V-12 coaxial loudspeakers, two to either side of the altar and two located halfway down either side of the room. The front pair of loudspeakers is delayed roughly 8 ms to encourage the precedence effect and the rear pair is delayed so as to be acoustically coincident with the front pair. A fifth V-12 faces the left transept, and the cry room is serviced by a pair of Tannoy V-8 systems. An additional pair of V-8 loudspeakers provides coverage to the altar area as seen in the photo of Figure 7.3.

Figure 7.3 Altar in St. Rose Catholic Church.

The acoustical environment of this church is well balanced; not too reverberant for speech but sufficiently so for traditional organ and choral music. Contemporary music suffers from a lack of definition due to the acoustics but there is only one contemporary-style Mass per week, so there is no great interest in making any significant changes to the room.

Report by Don Washburn

Case study

St. Dominic Catholic Church

Electro-Acoustic Consultants: The Audio Bug Inc.

St. Dominic Catholic Church (Figure 7.4) is located near Miami Airport. The current structure was completed in 1981. The original sound system was totally unacceptable! Cheap column speakers, one to either side of the altar, provided zero intelligibility.

Our firm was involved in the only upgrade. The system now consists of three Electro Voice (EV) HR-6040A horns fitted with DH1A drivers. Two 15-inch Community Sound boxes are located one to either side of the

Figure 7.4 St. Dominic Catholic Church, Miami.

Figure 7.5 St. Dominic Interior.

center horn. They're spaced as follows: one on the center-line of the church with the remaining two 60 degrees left and right of the center device.

As shown in Figure 7.5, the room is a large 170-degree fan shape, and virtually all surfaces are hard. The ceiling is low toward the back perimeter of the room, rising to over 30 feet above the altar. For such a small floor plan, the internal volume of the room is enormous. T60 exceeds two seconds throughout the mid-band.

They use primarily contemporary music because the congregation is 99 percent Latino. Drums and other percussion are very difficult to use in such a reverberant space, however they seem content with things as they are so we dropped the issue of acoustical treatment after a few discussions.

Report by Don Washburn

End Notes

1. Refer to Marshal, M. (1969). *The Gutenberg Galaxy: The Making of Typographic Man*, New York: Signet; and the work of Walter Ong, especially *Orality and Literacy: The Technologizing of the Word*, New Accents, Ed. Terence Hawkes, New York: Methuen, 1988.
2. Webb, S. H. (2004). *The Divine Voice*. Brazos Press.
3. Ibid., p. 37.
4. Ibid., p. 40.

EVANGELICAL WORSHIP

8

THE EVANGELICAL STYLE OF WORSHIP
So That All May Hear

CHAPTER OUTLINE

Introduction

The Evangelical worship style is identified by the *centrality* of the desire to preach or to proclaim the Gospel. The word *evangelical** is a compound word combining two Greek words, *eu* meaning good and *angelion* meaning message. The word Gospel is also a compound word from old English, meaning good news. These are the churches that focus on the message that salvation from eternal damnation is possible thorough Christ. This is the "good news" or Gospel. On the surface, most every Christian church will insist that dissemination of the Gospel is at the heart of what they do. But when we examine the Sunday morning service and how the church building is used we can see significant differences.

*Please note that in this chapter the term "evangelical" will be used to describe churches that employ a worship *style* and emphasis, which we are calling evangelical. Unless explicitly stated otherwise, the term evangelical written with lowercase "e" will refer to the worship *style* and will not necessarily refer to those churches that are evangelical in *theology*. When we refer to those churches that belong to the greater Evangelical movement or are Evangelical in belief we will use uppercase. It may be possible for a church with Evangelical theology to adopt a different form or style of worship.

Sound of Worship. DOI: 10.1016/B978-0-240-81339-4.00008-9

THE GOSPEL

The Gospel, or Good News, is the heart of all true Christianity. In a nutshell, the Good News is that although all humanity has sinned and grieved God, God has provided a way for reconciliation and salvation resulting in eternal life in heaven. This provision is through the death and resurrection of Jesus Christ, who although sinless Himself made Himself a perfect sacrifice for the sins of the world. This core belief is held with some nuance of difference by all who call themselves Christian. The belief in this Gospel along with the beliefs outlined in the Nicene Creed (see Chapter 3) constitute the essential dogma of the Christian faith.

To best understand the evangelical style, contrast it with the Celebratory. The Celebratory style focuses on the Mass, and in particular the Sacrament of the Eucharist, whereas in the evangelical style, the *focus* is on preaching the Gospel so that nonbelievers will believe and be saved.

It is in these churches that the Calvinist and Arminian doctrines play out most sharply. Those churches that are evangelical in style and Calvinist in doctrine and praxis will see the need for evangelism in the context of God's omniscience and omnipotence and the irresistible grace that God bestows upon the elect. And while in Calvinist thought the church will not be held accountable for not spreading the word effectively, the church is one of the vehicles that God used in the process of drawing

His elect unto Himself. The Calvinist will emphasize the importance of right thinking and correct theology as being important to a vibrant and true Christian experience, and in some circles, salvation itself. This right thinking and correct theology will come from the scriptures, but the primary place the scripture is interpreted is the pulpit.

The Arminian who employs an evangelical worship style has a lot more at stake. Since in Arminian thought the free will of man is emphasized, the efficacy of the church service and the preaching in particular is critical. If the preacher is not effective in getting sinners to the point where they acknowledge their need and seek salvation, the sinner will remain in his or her unsaved state; should he or she die in that state, the sinner will face eternal damnation.

Typical Evangelical Worship Service

Although there is considerable variability in the evangelical style of worship, the vast majority of churches worshipping in this style will employ these elements.

1. Prayer

   The Protestant understanding of prayer is straightforward. There are many different reasons for praying, but basically in the Protestant mind prayer is nothing more than talking with God. Unlike in the Catholic tradition, the members of the Trinity are the only legitimate recipients of prayer. There will be prayers of invocation where God is invited to be present in the service. There will be prayers of dedication or consecration—for example, when the "offering" is collected the money and gifts will be consecrated to God by prayer. There will be prayer of entreaty where requests are made of God. These may be personal or corporate requests. There will be prayers of blessing in which God is asked for special favor to be bestowed upon individuals or organizations.

2. Music

   There are different types of music used in the evangelical church. Much of the understanding of how music can and should be used in worship can be traced back to the revivalists of the nineteenth century. It was during this period that a new type of church music was created—the gospel song. These songs sometimes recounted the experiences of those already converted, but they also were designed to make the conversion experience attractive to those not yet converted.[1]

Stylistically, music can vary widely from contemporary to traditional. The message in this music is very important. The lyrics must be theologically correct and must reflect some element of the Christian experience. A primary purpose for music in the evangelical service is to offer praise to God for His salvation. Contemporary in this context generally means music composed in the last decade and utilizing guitars, keyboards, and possibly drums. Generally it won't be hard rock. On the other hand, traditional music is often many decades old, and may even be music that has been in use by the Church for centuries. Some of Martin Luther's hymns are very popular in the Evangelical church more than 400 years after Luther's death. Traditional music will generally be accompanied by piano, organ, or both.

Whether contemporary or traditional, the music will generally focus on what God has done for the individual, and the individual's praise and thanksgiving to God. The music then reinforces the correct theology and praxis that are so important to this style of worship.

In addition to the corporate music where the whole congregation is singing often from printed hymnbooks, there is also music that will accompany other activities, most often the receiving of the offering (see housekeeping, later). This is called the offertory and is usually performed by a musician at the piano or organ, sometimes with a vocal accompanist. There is often "special" music, which is performed not for the entertainment of the congregation, but for their edification. If the church has a choir they may perform, but it may also be a soloist singing along with a live keyboard player or with a pre-recorded track.

3. Scripture Reading

   In the evangelical style of worship the reading of the scripture plays a very important role, probably second only to the preaching. Scripture may be read by lay people or by the pastor. Sometimes scripture is read responsively, where the pastor will read a line and the congregation will read a response in unison. Most churches that employ the evangelical style will utilize hymn books that have extensive responsive readings printed in the back of the book. Scripture is viewed as the full revelation of God and the public reading of the scripture sets the context for the preaching that generally follows. As most of the Protestant Church is unaware of or ignores the traditional Church calendar followed by those churches who are more Celebratory (Catholic, Orthodox, Episcopal, etc.) there is no prescribed reading that all churches will follow on a given Sunday. Nor is there an expectation of reading from different parts of the Bible like in the Celebratory service. The leaders of the denomination may suggest a reading and sermon topic but the Evangelical church is largely a product of fiercely independent thinkers, from the founding fathers of the reformation to the modern "church planters." These pastors may impose a top-down structure in their own congregations but will generally resist being influenced by a denominational or other clerical authority. This is clearly a residual effect of the reformation where the authority of the Church was replaced with the authority of the Scripture and the hierarchical structure of church governance replaced with a single person in charge and the "priesthood of believers." Unlike in the Celebratory tradition, there is no special emphasis placed on the Gospels, or the words of Christ.

4. Preaching

   The sermon is the high point and the center of focus of the evangelical worship style. It can last from 30 minutes to over an hour. The sermon may be *topical*, in that the preaching may be about a topic or theme with numerous scripture references to illustrate and support each point and give the sermon authority. It may be *exegetical*, in which a biblical text is explicated and expanded on again with numerous references to other scripture to add support and an explanation of the focus text. In Evangelical churches the scripture is the final authority for everything, hearkening back to Luther's "solo scriptura." It will trump any church teaching, any sort of tradition, and all personal revelation. The preaching serves two distinct purposes. First, it is designed to present the Gospel to anyone in the congregation who either may have never heard the Gospel message before, or having heard it never made a decision of faith and accepted the salvation offered by Christ. Second, the sermon is designed to challenge the faithful to live a better or fuller Christian life. In his paper, *German Mass and Order of Divine Service* (Jan 1526), Luther wrote of the need of a new form of the Mass: "But, above all, the Order is for the simple and for the young folk who must daily be exercised in the Scripture and God's Word, to the end that they may become conversant with Scripture and expert in its use, ready and skillful in giving an answer for their faith, and able in time to teach others and aid in the advancement of the kingdom of Christ."[2] This is the primary vehicle by which the Christian receives teaching and encouragement in his religious life. But why is preaching so important to the Evangelical? Partly it is due to history. Luther and Calvin both stressed the importance and indeed primacy of preaching. Evangelicals believe that it is also stressed in the scripture. Romans 10:17 reads, "So faith comes from hearing and hearing by the word of God."[3] John White points out in his book *Protestant Worship and Church Architecture*, "...Protestants see preaching as dependent upon the power of God. It is not intended to be an inspiring talk or even an exciting challenge to action presented by the minister on his own authority....(it) is a means by which the power of God is made present in the midst of His people."[4]

5. Invitation

A key feature of the evangelical style of worship will be the "invitation" or "altar call." This will normally occur at the end of the preaching time and is an invitation for anyone who feels that they would like to be "saved" or "accept Jesus as Savior" to come forward to the "altar," generally a space just in front of the pulpit. The penitent would then be counseled by the Pastor or some other qualified attendant, and generally after a few questions they would be led in a prayer. The *theological* Evangelical would hold that when the penitent says this prayer, he or she is then saved. If the church holds a Calvinist perspective, they would teach that this act is irrevocable and permanent. Of course the more Arminian would teach that the prayer was efficacious—the person is saved—but the salvation could be undone. This practice is a relatively recent development. The altar call was used by Wesley and Whitefield in eighteenth century English revivals, but it was Charles Finney (1792–1875) who honed it into an art form using music and counting on the "herd mentality" to increase the numbers of penitents who came forward to experience the transformation of salvation. Finney was determined to " use whatever means were available to him to bring the unrepentant to God through a conversion experience. Finney scorned traditional forms of worship unless they were efficacious in producing conversions."[5] Finney and others developed new strategies called "New Measures" to reach the lost and as a result remade Protestant worship. These new measures included "public confession of faith, the altar call and the individual struggle with the soul which proceeded publicly upon an 'anxious bench' facing the congregation."[6]

6. Other Elements

There are a number of other elements to the evangelical style of worship that bear mentioning. In each service or meeting there will almost always be a collection of the offering or monetary gifts that are used to support the ministry of the Church. This is often done by ushers passing offering plates or baskets around until all have had the opportunity to give. Participation is voluntary.

TITHES

The concept of giving back to God one tenth (a tithe) of all that one earns dates back to ancient times. In the Old Testament it is commanded by God as a way to worship and honor God and as a way for the priests to live. In the New Testament it lives on and has been passed down through the generations and various iterations of church and lives on in virtually every form of Christian church. A tithe is considered a minimum amount, and most churches expect that members will give more than a tithe. In the nineteenth century, many American churches charged a pew fee to those in attendance. In the modern American Church the "pew fee" is gone and participation in giving to the Church is totally voluntary.

In each service there will be a time set aside for announcements of church activities and to inform members about specific needs that individuals might have. These will often be presented in the form of "prayer requests," inviting the faithful to pray on behalf of those in need.

Evangelical churches that embrace a Baptist theology will place special emphasis on baptism. There will often be a baptismal pool in the sanctuary and it will vie for architectural prominence with the pulpit. Baptisms will occur in regularly scheduled services.

The practice of participating in the "Lord's Supper" (also known as communion) in the Evangelical church is widely varied. Whereas they will all agree that the "Lord's Supper" is symbolic only and something that imparts no particular mystical merit, the frequency of participation depends on the preference of the particular church. Some will serve communion every Sunday, others once a month, others once a quarter, others once a year. What started out as bread and wine as part of a meal has devolved into thin wafers and grape juice served to the congregation in trays. Most Evangelical churches will ask that participation in communion be limited to those believers who are baptized.

Typical Architecture of Evangelical Style Church

We will address the acoustics of the Evangelical church in a separate chapter, but we want to briefly look at the features of the architecture of the Evangelical church that are unique to it and that distinguish it from the other types of churches. Architects still like to debate the merits of the oft-cited Louis Sullivan edict that "form {ever} follows function." In the Evangelical church, the form of the church building will virtually always follow the function of the space. From the exterior, the typical Evangelical church will be rectangular and of moderate aspect ratio. There may be stained glass, but most often it will be abstract and not representative. There will generally be a steeple, but that will be the extent of the symbolism. In this style of church, there is virtually no place for symbolism. Indeed, other than music, there is relatively little space for art at all. The church building is not viewed as a sacred or holy space at all. It is the activity that makes the space sacred. The Evangelical will not think of the pulpit or the communion table or the baptismal as being sacred objects. The preaching of the word, however, from that pulpit is a sacred act, as is the participation in the communion or in performing baptism.

When you walk into an Evangelical church (Figure 8.1), you notice immediately that the most prominent feature is the pulpit. The pulpit will be in the center of a stage area that is almost always raised up above the congregation. This is the center of focus and the place from which everything emanates. The pulpit may not be a sacred object in and of itself but it does have a sacred *function*.

In a physical sense it simply is a place that holds up the preacher's notes and maybe a microphone stand. It is far more than this on a spiritual and (ironically) on a symbolic level. The Evangelical church along with most of Protestantism rejects symbolism, yet in the pulpit we find a powerful symbol of authority and truth. It is the starting and ending point

Figure 8.1 Interior of a Typical Small Country Baptist Church.

for the Evangelical church and rightly occupies the center of visual focus. On the stage with the pulpit there will often be conspicuous seating for the pastor and other leaders of the worship service.

There is almost always a table directly in front of and below the pulpit. This table is a sort of a vestigial altar combined with a common ta communion. ble. As an altar, it is symbolic of sacrifice, or man's service to God, and it is here on this altar of sorts that the tithes and offerings are placed upon collection. As a table it represents God's gifts to man and it is from this table that communion is served. It is not by accident that the pulpit is above the altar/table. Everything is subjugated to the Word, even the communion. In some Evangelical churches there will be a cross. The cross will be prominent in the room, maybe even above the pulpit as it is the symbol of salvation, the core concept of the Gospel. The cross will always be empty. This is in stark contrast to the crucifix seen in Catholic and other Celebratory churches.

CROSS OR CRUCIFIX?

A crucifix is a cross that includes the figure of Christ in the process of being executed. This is a point of controversy between Catholics and Protestants. Those who use crucifixes often cite the verse, "we preach Christ crucified: a stumbling block to the Jews and foolishness to the Gentiles" (1 Corinthians 1:23). The emphasis is on the suffering and the sacrifice. Those who use plain crosses will point out that Jesus did not stay on the cross and that he is risen from the dead. They will point out that many were crucified, but only Jesus triumphed over death and that is why he is preached and worshipped. They point to the Catholics especially and accuse them of leaving Jesus on the cross. The Catholics for their part would point out that the empty cross can "cheapen" the message that Christ indeed did suffer and that salvation had a terrible price.

If the Evangelical church is from a Baptist tradition, there will be a baptismal. This will not be a simple or small font holding consecrated or holy water. The Baptist and most Evangelicals will insist on complete immersion in the water, not a mere sprinkling upon the head. The baptismal will be located strategically and be constructed at least partially of glass or other transparent material so that everyone in the congregation will be able to see the convert descending into and emerging from the water.

For congregational seating, pews are most common. They are usually set up parallel to the front of the church, situated so that everyone can easily see the pulpit. There will not be kneelers as kneeling by the congregation has virtually disappeared from the Evangelical church. The pews will accommodate copies of the Bible and also hymn books placed in racks in the pew-backs. There will also be miniature cup holders built into the back of the pews to accommodate the miniature glasses used for communion.

Most Evangelical churches will have a choir loft of some sort, generally in the front of the church, either behind the pulpit or off to one side. There will also be provision for a variety of musicians ranging from a piano and electronic organ to a full blown mini orchestra in larger churches. Some will use a praise and worship "band" in what would be called the contemporary worship service. It is important to reiterate that music in the Evangelical service must support the overall mission of preaching the gospel.

Historical Evolution of the Evangelical Worship Style

It may surprise advocates of this worship style that it is a relatively modern development. Most of what happens in a evangelical style church has roots in the revivalist movement of the nineteenth century. It was during this so-called "Second Great Awakening" that for much of the Protestant church, worship began to change its focus from a service or "work" that was performed for the glory and benefit of God to an activity that focused on the individual and his or her need for salvation. The phrase "I did not get much out of the service" is an odd construct! Indeed the use of the word "service" as in worship service or morning service reflects an earlier understanding of what worship was: an act focused on God rather than on me.[7] However, to fully appreciate the intensity of devotion to the Word, we must look at the Reformation.

End Notes

1. White, J. F. *Protestant Worship and Church Architecture* (p. 8). Eugene Oregon: WIPF and Stock Publishers.
2. Kidd, B. J. (Ed.), (1911). *Documents Illustrative of the Continental Reformation.* Oxford Press. Accessed online at http://www.iclnet.org/pub/resources/text/ wittenberg/luther/germnmass-order.txt, 8/25/09.
3. Romans, Chapter 10 verse 17, New American Standard Bible.
4. White, J. F. *Protestant Worship and Church Architecture* (p. 37). Eugene Oregon: WIPF and Stock Publishers.
5. Ibid., p. 8.
6. Kilde, J. H. (2002). *When Church Became Theatre* (p. 22). Oxford University Press.
7. Ibid., p. 17.

9

THE SIXTEENTH CENTURY REFORMERS: LUTHER, ZWINGLI, AND CALVIN

CHAPTER OUTLINE

To the Roman Catholics it was a rebellion. To Protestants, it was their birth and a reformation in the truest sense of the word. To society it was a revolution. To the Orthodox it was a Roman Catholic problem and had little relevance. No matter what position you hold, the Protestant Reformation certainly changed the Christian Church. If we are to understand and therefore serve the church who worships in the evangelical style we need to understand the reformation. To that end, we will look at the societal forces that made it possible.

The Renaissance

The so-called Renaissance or rebirth was tearing through Europe with a vengeance, fueled in part by the invention of the printing press in the 1450s. The term *Renaissance* implies that a profound change or transition was taking place. The nature of this transformation holds the key to understanding the Reformation.

Sound of Worship. DOI: 10.1016/B978-0-240-81339-4.00009-0

It is admittedly irresponsible to suggest that a cultural movement like the renaissance can be reduced to a simple formula. There are, however, at least three things that were reborn.[1] First, classical scholarship, or the revival of learning involved the rediscovery of Latin and Greek manuscripts and the study and publication of these works. During the renaissance, it was fashionable to read anything Greek. Second was the rebirth of the notion of the free individual. As the intelligentsia read the classics they rediscovered unfettered thinking. In the works of Plato and Aristotle "they heard a call to a kind of freedom that they had not known."[2] Third, intellectual curiosity was reborn. This was probably the most important of the three. When freedom from conformity to the establishment was combined with curiosity it is not surprising that the world changed.

During the Renaissance the arts flourished as creative individuals explored new secular themes that had been off limits. Others turned their creative energies to the exploration of basic problems, and a philosophy that did not find its roots in theology emerged. Some of the great scientific minds, most notably Nicolaus Copernicus (1473–1543), lived during this time. Copernicus was a priest and mathematician, and is credited with being the first to propose that the sun was at the center and the planets revolved around it. During this time of transition it was very dangerous indeed to think let alone write such thoughts and Copernicus was reluctant to publish his work.

The Church was not immune to the societal reawakening that swirled around it. The Roman Catholic church became an important patron of the arts and amassed important collections of art and literature. But the old adage about power corrupting and absolute power corrupting absolutely was in evidence in the Roman Catholic church. According to McBrien's *Lives of the Popes*, there were four Popes who ruled the Church between 1492 and 1521; Innocent VIII, Alexander VI, Julius II, and Leo X. All four made McBrien's list of the "worst of the worst," with Alexander VI (1431–1503, Pope from 1492–1503) accorded the dubious honor of being called the "most notorious Pope in history."[3] Roman

Catholic authors admit that this was not the finest era in the history of the Church.

Politically, by the fifteenth century, the merging of Church and state throughout Europe could not have been more complete. The corruption and excess that were rife in the Church were also those of the state. The religious leaders were also the civil leaders. From the point of view of the peasants, the state dictated the terms of their physical existence and the Church dictated the terms of their afterlife. There was no escape from this two-headed monster.

Early Protestant Movement

It may come as a surprise that Martin Luther was not the first to protest against the Roman Catholic church and to try to bring reform. In the late fourteenth century a Czech priest named Jan Hus (or John Huss, c. 1369–1415) began to question the Church. Profoundly influenced by the writings of John Wycliffe (see text box below, *John Wycliffe*), Hus wanted to bring the Church back to what he felt was a more correct position. The Church in Bohemia had been founded in the ninth century by two missionaries from the Western (Orthodox) church.[6] These Western missionaries had

JOHN WYCLIFFE

No account of the Reformation would be complete without some mention of John Wycliffe (1320–1384). Wycliffe was an Oxford scholar and theologian, and the first to translate the Bible into English. His source was the Latin Vulgate as the earlier Greek texts were not available to him. Wycliffe is rightly remembered as the translator but he also was responsible for sowing the seeds that would later grow into full blown reformation. He took very unpopular positions on many of the teachings of the Roman Catholic church of his day, including questioning the validity of the papacy itself. It is not fully known who were Wycliffe's primary influences, but it is clear that like the other great reformers who followed him centuries later, Wycliffe based his claims and positions on his study of the Bible. He is quoted as saying, "Even though there were a hundred popes and though every mendicant monk were a cardinal, they would be entitled to confidence only in so far as they accorded with the Bible."[4] And again, "The Church is the totality of those who are predestined to blessedness. It includes the Church triumphant in heaven... and the Church militant or men on earth. No one who is eternally lost has part in it. There is one universal Church, and outside of it there is no salvation. Its head is Christ. No pope may say that he is the head, for he can not say that he is elect or even a member of the Church."[5] It comes as no surprise that Wycliffe was not held in high esteem by the papacy, and 44 years after Wycliffe died of natural causes, the Pope declared him a heretic and ordered that his body be exhumed, and his bones ground up and scattered in the river. A number of Wycliffe's hand written manuscripts remain.

instituted a national liturgy and as was the practice of the Western church, they had translated the Bible into the common tongue. However by the fourteenth century, Bohemia and Moravia had come under the jurisdiction of Rome and many of the Western aspects of the church were eliminated. Hus was an outspoken critic of the Roman church and had wide popular support. He did not, however, have the support of the Roman church and on July 6, 1415 he was burned at the stake. Some 40 years later, about 60 years before Martin Luther began his reformation efforts, his supporters founded the Moravian church or *Unitas Fratrum*. This new church addressed many of the evils and excesses of the Roman church and also to some degree broke away from its Orthodox roots. They emphasized the role of scripture and, most importantly, they supported one of Hus's primary teachings, justification by faith alone. This doctrine sets the Moravian church clearly outside the Roman Catholic or universal church and thus it can be rightly called the first true *Protestant* church. The

Moravian church still exists today, and although small in number, they have been remarkably effective in their outreach or missionary work around the world.[7]

Martin Luther

Onto this stage walks a Roman Catholic priest, Martin Luther (b. 1483; Figure 9.1). Luther was born in the Saxon mining town of Eisleben. His parents gave him the best education they could afford, hoping that he would become a lawyer and aspire to a lucrative career. To his parents' dismay he joined the priesthood and became a monk. Luther was a dedicated and brilliant if restless monk. The leader of the monastery encouraged him to take an advanced degree in theology so he could teach at the University of Wittenberg.

Figure 9.1 Martin Luther (1483–1546).

Luther in his own words discusses his transformation, which came largely through his reading and meditating on the scriptures:

Meanwhile in that same year, 1519, I had begun interpreting the Psalms once again. I felt confident that I was now more experienced, since I had dealt in university courses with St. Paul's Letters to the Romans, to the Galatians, and the Letter to the Hebrews. I had conceived a burning desire to understand what Paul meant in his Letter to the Romans …

I meditated night and day on those words until at last, by the mercy of God, I paid attention to their context: "The justice of God is revealed in it, as it is written: 'The just person lives by faith.'" I began to understand that in this verse the justice of God is that by which the just person lives by a gift of God, that is by faith. I began to understand that this verse means that the justice of God is revealed through the Gospel, but it is a passive justice, i.e., that by which the merciful God justifies us by faith, as it is written: "The just person lives by faith." All at once I felt that I had been born again and entered into paradise itself through open gates. Immediately I saw the whole Scripture in a different light. I ran through the Scriptures from memory and found that other terms had analogous meanings, e.g., the work of God, that is, what God works in us; the power of God, by which he makes us powerful; the wisdom of God, by which he makes us wise; the strength of God, the salvation of God, the glory of God.[8]

This realization that it was by faith alone that we find God was of course heresy from the point of view of the Roman Catholic church. Roman Catholic doctrine taught (and still teaches) an ecclesiastical authority that is on a par with the Scripture. This insistence that the scripture alone is the final authority became known as "Solo Scriptura," and to this day it is one of the beliefs that separates the Roman Catholic church from the Protestant churches.

INDULGENCES

An indulgence was something that one could buy from the Church, which granted the buyer a certain kind of remission—not from the *guilt* of sins but from the *penalty* of sin. The idea was that if you were guilty of some moral lapse or sin, your first obligation was to seek forgiveness from God through the Church via the process of confession and penance. This would absolve you from the guilt of the sin, but the penalty would still remain. You could pay a fee to the Church and you would be spared the consequences of the sin in the afterlife. This practice was grossly abused during Luther's time. It is said that a great deal of the money that went in to the construction of St. Peter's Basilica in Rome was money gathered through the sale of indulgences. Luther called the indulgences " pious frauds of the faithful."[9] The Roman Catholic church still defends the practice but acknowledges that abuses have occurred.[10]

In 1517 while Leo X was Pope, Martin Luther wrote *In Disputation of Doctor Martin Luther on the Power and Efficacy of Indulgences*, also known as the *95 Theses*. In a bold move, he nailed them to the door of the castle church in Wittenberg, Germany. The act of nailing them wasn't such a big deal—this was the sixteenth century version of posting something on the Internet. But going public with his beliefs was. The document was really 95 points that Luther wished to debate, but it became the symbol of a new break with Rome, the beginning of the Protestant Reformation. Luther continued to write, his work growing more and more inflammatory. In June of 1520 Pope Leo X issued a proclamation that resulted in the excommunication of Luther. Luther's books were burned in Cologne. Luther, not easily intimidated, publicly burned the Pope's decree. In April of 1521, he was called before the emperor to defend his writings and ostensibly recant at least some of them. His trial was to be held in the German town of Worms and is known by the unappetizing title, the Diet (assembly) of Worms. At his trial, he boldly challenged the Emperor, and of course by extension the Pope, to prove him wrong *on the basis of Scripture alone* (solo scriptura). In addition to being excommunicated from the Church by the Pope, he was now branded an outlaw by the Emperor Charles.

Protestant readers might benefit by seeing some of the Roman Catholic church's reply to Luther, part of which reads, " '...if it were granted that whoever contradicts the councils and common understanding of the Church must be overcome by Scripture passages, we will have nothing that is certain or decided.' Luther's conscience was captive to the Word of God. But the imperial court was quick to ask a disturbing and discerning question—what if everyone simply followed his or her own conscience? The end result was obvious—'we will have nothing certain'."[11] The Roman Catholic church legitimately asks the twenty-first century Protestants, where has *Solo Scriptura* gotten you?

The Roman Catholic church remains a reasonably united organization, with its own divisions and disagreements to be sure, but the Protestant church is not a church at all but rather a collection of strongly independent organizations with little or no connection between them, united only in what they are *not* (not Catholic) rather than by sharing any common identity or ideology.

Luther was a prolific writer and by 1534 he had translated the entire Bible into German. This was almost as radical as publicly challenging the Papal authority. Up to this time the Bible was available only in Latin and meant to be read only by the clergy.

Even though most people in Luther's time still were illiterate, the literacy revolution started by Jonahs Gutenberg 100 years or so earlier was sweeping the land. Luther, by translating the Bible into the common language of the people, was sending a message to Rome that individuals did not need the agency of the Church to give them limited access to the Word of God. They could read it themselves and ultimately *interpret* it however they pleased. This was a serious undermining of the authority of the Church.

From his point of view Luther remained faithful to the Roman Catholic church, which he passionately wanted to reform. This reform movement caught on rapidly in Germany to the extent that in 1526 in an assembly in Speier, the followers of Luther were able to get the Diet to agree that the ruler of each state should be free to decide on the correct faith.[12] This freedom was short lived however, for in 1529 there was a second Diet at Speier, which overturned the earlier policy and made the Roman Catholic church once again the only legal one. Luther's followers read a *protestation*, and from then on were known as protestants.[13] Martin Luther died a natural death—a miracle in itself—in 1546. During his lifetime there was no organized Lutheran church per se, but after his death his followers organized and the Lutheran church is still an important part of the greater Protestant church.

Luther is often criticized in modern times for his strong anti-Semitic statements. Like every important historical figure, Luther is not one dimensional. He did some admirable things and he did things which were certainly regrettable and downright offensive. His courage to stand against the Roman Catholic church took a great deal of personal courage and depth of conviction. But at the same time his anti-Semitic writings, although not unusual for his time, should be condemned for what they are.

Ulrich Zwingli (1484–1531)

Luther is rightly credited with the beginning of the Protestant movement. But to fully understand the Protestant church today, we must look at a few other historical events and individuals.

According to Hutchinson, "The Protestant Reformation was not one movement that later divided. It was at least four movements which never united."[14] One example is that of Ulrich Zwingli (Figure 9.2), the leader of the reformation movement in the German cantons of Switzerland. A study of his life would reveal a remarkable parallel to Luther. He was born in Switzerland, a child of a family from the almost nonexistent middle class who was educated beyond what could be considered the norm. Although he apparently did not ever finish seminary, he became a priest in the Roman Catholic church. In December of 1518, he was appointed priest of a large church in Zurich. In Zurich he, like Luther in Germany, became concerned with the level of corruption in the Church. His preaching and teaching gradually strayed further and further away from the teachings of Rome. Like Luther he claimed that he based all of his positions solely on the Scriptures. In 1519, a special envoi from Rome arrived in Zurich to sell special indulgences for contributors to the construction of the great St. Peter's basilica in Rome. This was over a year after Luther put up his *95 Theses*. Zwingli pushed back rather forcefully, complaining that the common people were not being given full disclosure as to the reason behind the sale of the indulgences. Surprisingly, Rome backed down, probably trying to contain the uproar started by Luther in Germany a year earlier. There is not much evidence of a lot of contact between Luther and Zwingli, but there was some. They actually agreed on many points of theology, except one very important one. Luther held the Roman Catholic view known as transubstantiation, teaching that Jesus was present in the bread and wine of the Eucharist in a miraculous way. Zwingli believed that the Eucharist was a symbolic act only and therefore had no miraculous or supernatural aspect at all. Luther refused to support Zwingli, and the two churches developed separately with Zwingli ultimately joining forces with John Calvin. Not surprisingly there were many significant differences among the Protestant ranks. Just how does one derive authority from scripture? Is it the only authority? As an example, Zwingli had the organ removed from his church because he could find no scripture that *called for* the use of organ music in Christian worship. Luther,

Figure 9.2 Ulrich Zwingli (1484–1531).

on the other hand, embraced music of all forms as he could find no scriptural *prohibition* of music in worship.[15] Zwingli is at least partially responsible for the difference in the way Protestant churches and Roman Catholic churches look today. Zwingli became convinced that statues of saints and other holy trappings had no basis in scripture and should be removed from the church building. Zwingli was one of the first to suggest that the Mass itself be abolished as it also had no basis in scripture. Although he clearly had significant impact on the development of Protestant thinking, outside of Zurich he is not credited with starting or founding any churches. Zwingli was a fascinating complex individual. He was a free thinker whose ideas helped to shape Protestant theology. He was a politician who fought for reform in the Swiss Confederation, and he was passionate about his beloved adopted city, Zurich. He died in a civil war between cantons known as the Second Kappel War (1529–1531), not as a martyr for his faith but as a patriot for his city.

John Calvin (1509–1564)

We turn our attention to a third important figure in the Reformation, John Calvin (Figure 9.3). French born Calvin was very different from Luther and Zwingli. Calvin was born into the professional class and received an education very different from that of Luther. Calvin had a legal and humanist training. He was raised in the Roman Catholic church but by the time he was a young adult the Protestant movement had already spread through much of Europe and he converted to a more Protestant viewpoint. When violence broke out in his native France aimed primarily at those who were or sympathized with Protestants, he moved to Basel, Switzerland, where he wrote the first edition of perhaps his greatest work at the age of 26. *The Institutes of Christian Religion* was a systematic theology aimed at defending the Protestants of France who were suffering for their beliefs. He would return to this work frequently throughout his life, publishing numerous versions and revisions. Calvin was approached by some who were trying to reform the Church in Geneva. Geneva was not ready for

Figure 9.3 John Calvin (1509–1564).

these radical ideas and Calvin and others were expelled from the city. He moved to Strasbourg, which at the time was a sort of city/state of the Holy Roman Empire and had become a safe haven for Protestants. He spent a number of years there working with French refugees. Calvin never lost his zeal for the reformation of the Church in Geneva and eventually he was invited to return to Geneva and pastor the Church there. He instituted sweeping reforms and was a prolific writer and preacher. In 1542 Calvin published his Catechism of the Church in Geneva. This catechism, a sort of theological handbook, was based loosely on Luther's earlier work, with some important changes in emphasis. Calvin was developing a theology and a system of church government (ecclesiology) that would have significant differences from Luther.

Calvin promoted a new form of church government, where Elders or Presbyters were elected into office and were the authority in a local body. His movement swept throughout French-speaking Switzerland and became the state church of Geneva. Of course Calvin had his own enemies, both from the Roman Catholic church and from other would-be reformers. In medieval minds, right thinking was rewarded. Wrong thinking was punishable by death. It may seem curious to modern readers that the reformers who were the radicals of their time and often under sentence of death themselves were so intolerant of others. By 1546 Calvin and his followers had executed 58 and exiled 76 individuals who disagreed with the position of what was now known as the Reformed church. Even though Calvin was directly or indirectly responsible for the executions, he frequently was arguing for the more humane forms of execution—like beheading rather than burning at the stake. Calvin managed to survive many attacks on his teachings at Geneva and spent the last decade or so of his life in relative peace. He was recognized as a reformer on a par with but distinct from Martin Luther. When the debates between Zwingli and Luther broke out over the issue of transubstantiation, Calvin's response seemed to Luther to place Calvin squarely in support of Zwingli. In spite of this rift with Luther and the rather heated exchanges between the Lutheran and Reformed churches, Calvin remained concerned at the lack of unity among the reformers. He signed an agreement with the Church in Zurich and attempted to build a bridge with the emerging Church of England, which was breaking away from Rome for its own reasons.

Calvin died of natural causes in 1564 shortly after finishing a last revision of his *Institutes*.

TULIP

In 1618, some 50 years after his death, the followers of Calvin, now called the Reformed church, published a concise statement of their views. It is known today as the *Five Points of Calvinism*. It is also known as TULIP due to the first letters of the five points.

Total Depravity

The Reformed or Calvinist church taught that humans are born with what is called original sin due to the actions of Adam and Eve back in the garden. Due to this condition, human kind is not able to reach out to God. Faith is seen as a gift from God enabling humankind to seek for and find salvation.

Unconditional Election

Unconditional Election means that God has determined before the beginning of time who will be brought into salvation, and is based solely on His sovereign purpose. This is not based on any merit on the part of the "elected." God, as a result of his election, grants faith and the ability to seek Him to the elect.

Limited Atonement

The death and resurrection of Jesus, which is the ultimate means of salvation, is effective only for the elect.

Irresistible Grace

When God calls an individual who is elect, unto Himself, that call is irresistible. The elect has no choice but to respond to God with faith supplied by God. God's grace will always win out and result in the elect coming to faith in God.

Perseverance of the Saints
The elect, once drawn to God though the work of Jesus, are then saved forever and cannot through any act of their will, fall from the grace which holds them.[16]

Calvin's theology still has major influence in modern evangelical Christianity, especially in the United States. His form of church government lives on in the Presbyterian church and others, and the Reformed church traces its roots directly back to Calvin. It is appropriate to mention at this point the contribution of one Jacobus Arminius. Arminius (1560–1609) was a theologian who had been taught and mentored by Calvin's successor, but who came to significant disagreement with Calvin's theology. His followers were known as the Remonstrants and were declared to be heretics by the Reformed church. It was in response to the Remonstrants that the famous five points of Calvinism were developed. Virtually all the modern day so-called Protestant churches subscribe to one or the other of these two theological stands, some support elements of both but do not see themselves fully in either camp.

ARMINIAN THEOLOGY

These are the five points held by the Remonstrants or Arminians that were refuted by the Calvinists.

Free Will—Humanability
Human nature was seriously affected by the "fall" but it did not leave humankind totally depraved or in a state of spiritual depravity. Each person, sinner though they be, possesses a free will with which he or she can choose to respond to God's grace, or reject it and suffer the consequences.

Conditional Election

Election was determined by what God knew man would do. Since God is omniscient, and knows what choices would be made, God elected only those who God knew would respond to His call. Salvation is the result of man choosing God rather than God choosing man.

Universal Redemption

Christ died for *all*, but only those who believe in Him and respond to the call of salvation will be saved. Because of Christ's death, God is able to forgive sinners if they accept forgiveness.

Holy Spirit Can Be Resisted

The Holy Spirit calls all to come to salvation. However, since man has free will, this call can be and often is resisted. In this way God's grace does not "force" all to come to repentance.

Falling from Grace

Those who have responded to the Holy Spirit and are saved can lose their salvation if they sin and do not maintain the relationship with God. Note: Not all Arminians would agree with this point; many side with the Calvinist point of view of eternal salvation or security. [17]

In the New Testament, in the Book of John, we read that the "word became flesh and dwelt among us, and we beheld His glory."[18] In the Reformation the Word became the central focus and the final say on all matters of faith and practice. The changes that swept through mainland Europe left a permanent mark on both the religious and political landscapes. It is not surprising that similar events were taking place across the channel in England.

End Notes

1. Hutchinson, P., & Garrison, W. (1959). *20 Centuries of Christianity* (p. 185). New York: Harcourt Brace and Co.
2. Ibid.
3. McBrien, R. (1997). *Lives of the Popes* (p. 267). San Francisco: Harper.
4. http://www.john-wycliffe.com/history.html, accessed 9/09.
5. http://www.greatsite.com/timeline-english-bible-history/john-wycliffe.html, accessed 9/09.
6. http://www.moravian.org/history/
7. Ibid.
8. Translated by Brother Andrew Thornton, OSB, (c)1983 by Saint Anselm Abbey.
9. Catholic Encyclopedia: *Indulgences*
10. Ibid.
11. Noll, M. (1997). *Turning Points* (pp. 155, 156). Baker Books.
12. Cairns (1967). *Early Christianity through the Centuries* (p. 320). Zondervan.
13. Ibid.
14. Hutchinson, P., & Garrison, W. (1959). *20 Centuries of Christianity* (p. 201). New York: Harcourt Brace and Co.
15. Noll, M. (1997). *Turning Points* (p. 193). Baker Books.
16. http://www.the-highway.com/compare.html, A Brief Comparative Study of Arminianism and Calvinism, accessed 8/09.
17. Ibid.
18. John 1:14.

10

THE REFORMATION IN ENGLAND

CHAPTER OUTLINE

Background

While the continent was embroiled in the Reformation and Luther, Zwingli, Calvin, and Simons were exerting tremendous influence on religious and civil life, similar events were in play across the channel in England. The reformation in England gave rise ultimately to a number of churches, including most of those who currently use the evangelical worship style. These all were eventual offshoots of maybe the most famous outcome of the English reformation.

In 1521 Lutheran books appeared in England and were the subject of much debate at the famed White Horse Tavern in Cambridge.[1] These meetings apparently came to the attention of the King (Henry VIII) who, in 1525, wrote a book about the sacraments as a defense against the influence of Lutheranism.

However, seven years later in 1532, Henry VIII found himself in a bind. He was married to Catherine, the widow of his brother Arthur. Henry became convinced that this marriage was invalid and should be annulled. There also was the fact that Henry's consort Ann Boleyn was pregnant with his child, hopefully his son. Henry needed to have this offspring, if male, be the legitimate heir to the throne. Rome was not willing for a variety of reasons

Sound of Worship. DOI: 10.1016/B978-0-240-81339-4.00010-7

Figure 10.1 King Henry VIII.

to grant Henry the annulment he desired. Finally Henry came to the conclusion that if Rome would not consent to the divorce from Catherine and a marriage to Ann Boleyn, he would have to find someone who would. If the result of this marital situation was that the Church in England needed to break from the Roman church so be it. The archbishop of Canterbury was willing to give the King what he wanted and in 1533 the King married Ann. In 1534, the English parliament passed an Act of Supremacy, which severed the Church in England from Rome.

THE ACT OF SUPREMACY (1534)

Albeit, the King's Majesty justly and rightfully is and oweth to be the supreme head of the Church of England, and so is recognized by the clergy of this realm in their Convocations; yet nevertheless for corroboration and confirmation thereof, and for increase of virtue in Christ's religion within this realm of England, and to repress and extirp all errors, heresies and other enormities and abuses heretofore used in the same, be it enacted by authority of this present Parliament that the King our sovereign lord, his heirs and successors kings of this realm, shall be taken, accepted and reputed the only supreme head in earth of the Church of England called *Anglicana Ecclesia*, and shall have and enjoy annexed and united to the imperial crown of this realm as well the title and style thereof, as all honours, dignities, preeminences, jurisdictions, privileges, authorities, immunities, profits and commodities, to the said dignity of supreme head of the same Church belonging and appertaining. And that our said sovereign lord, his heirs and successors kings of this realm, shall have full power and authority from time to time to visit, repress, redress, reform, order, correct, restrain and amend all such errors, heresies, abuses, offenses, contempts and enormities, whatsoever they be, which by any manner spiritual authority or jurisdiction ought or may lawfully be reformed, repressed, ordered, redressed corrected, restrained or amended, most

> to the pleasure of Almighty God, the increase of virtue in Christ's religion, and for the conservation of the peace, unity and tranquility of this realm: any usage, custom, foreign laws, foreign authority, prescription or any other thing or things to the contrary hereof notwithstanding.[2]

Of course the story does not end here. A scant three years after marrying Ann, and dissolving ties to Rome, Henry VII had Ann beheaded on trumped up charges and married Jane Seymour. Seymour died following childbirth a few years later, but did produce Henry's only legitimate heir to the throne. There were three more marriages, one divorce, and one spousal execution before Henry died in 1547.

According to the Anglican Church's own timeline, "the History of the English reformation... is not very edifying...."[3] The Church of England (also known as the Anglican or Episcopal Church) was formed not because of some deep-seated theological or doctrinal differences. Unlike Luther, Zwingli, Calvin, and other leaders of the reformation in Europe, King Henry broke with Rome because he could. The Act of Supremacy was clearly part of a trend of "...establishing small scale alternatives to the Universal Catholic Church. This development forever changed the face of the Catholic church in the West."[4]

William Tyndale

During this time, William Tyndale (1494–1536) was working hard to translate the Bible into English. English-born Tyndale was educated at Oxford where he graduated with his master's degree at the age of 21.[5] He was fluent in Hebrew, Greek, Latin, French, Spanish, German, Italian, as well as his native English. Tyndale became convinced, as had the other reformers, that the Bible was an authority that superseded the authority of the Church. And, as society reaped the benefits of Gutenberg's invention, and more people learned how to read, Tyndale understood that giving the common man the ability to read the scriptures in his own

Figure 10.2 William Tyndale (1494–1536).

language was the surest way to undermine the authority of the Roman Catholic church, which he viewed to be corrupt. Once, a noted cleric said to Tyndale during a heated debate regarding authority, "We had better be without God's laws than the Pope's." Tyndale is reported to have responded to this bit of heresy, "I defy the Pope and all his laws: and, if God spares my life, I will cause the boy that drives the plow in England to know more of the Scriptures than the Pope himself."[6]

Certainly this sort of behavior did nothing to endear him to the King and the Church! Tyndale dedicated his life to making good on his outspoken "promise" and did in fact translate the Bible into English, completing work on the New Testament in 1526. The rest followed piecemeal over the next decade or so. This was not the first translation into English—that was accomplished by Wycliffe—but this was the first translation from the original languages. Initially there was considerable "push-back" to the very notion, let alone the presence of the scriptures in the vernacular. Many copies were gathered up and burned. Tyndale's writings would be his undoing, but not just the heresy of translating the Bible into English. In 1530 he published *The Practice of Prelates*, which was a very outspoken (even by Tyndale standards) criticism of Henry VIII, especially of his divorce of his first wife Catherine. Even though the title page states that the purpose of this work is to address the divorce, Tyndale does not actually deal with the divorce for 164 pages! This remarkable work starts with the New Testament account of the betrayal of Jesus and traces one conspiracy after another, through 1500 or so years of Church history, culminating with Henry VIII and the various co-conspirators of his time.[7] Translating the Bible may have been heresy but this was personal. Henry VIII asked the Emperor to arrest Tyndale (now living in Holland) and to charge him with treason. Tyndale was arrested in 1534 and was executed in 1536. His last words were, "Lord open the King of England's eyes."[8] His dying wish was granted as, in true Henry VIII fashion, three years later Henry published the Great Bible, the first authorized English Bible to be used in the newly formed Church of England. It was mostly

Tyndale's work. Tyndale's contribution to the greater Protestant church cannot be overstated. He had enormous influence on the development of English religious language.

One interesting development from this era is the emergence of music as a significant part of a church service. The use of music in the church was a topic of serious debate in the sixteenth century. The stand that the various groups would take regarding music, of course, was derived from their emerging theologies, and over the years had an impact on the architecture and therefore the acoustics of the Church. Luther, with some 37 hymns to his credit, loved music and encouraged the use of music and instruments in the Church. This arose partly out of his conviction that Church tradition was still important, just not the final word. Others like Calvin and especially Zwingli were very uneasy with anything that smacked of tradition and that could not be directly supported from scripture. Some of these churches prohibited the use of instruments in the service. They were more likely to set Biblical passages to music. The Church of today has certainly benefited from the musical diversity of the Reformation. The modern Mennonites, descendents from the early Anabaptist (see Chapter 17) who did not allow instruments in the service, still highly value a cappella singing. This author was privileged to hear over 10,000 Mennonites sing a cappella at a world conference. It was a truly awe-inspiring moment! According to Noll, "The flurry of Protestant hymn writing that burst into Europe alongside the early crises of the Reformation created unusual difficulties for the Roman Catholic church. So thoroughly was congregational song associated with Protestantism and so effective were the Protestants at putting hymns to use that leading figures in the Roman Catholic church briefly contemplated a ban on music in their services."[9] Fortunately for church music, the ban did not happen. The seventeenth century saw a tremendous outpouring of sacred music by some of the great composers of all time. Johann Sebastian Bach alone wrote 24 Masses!

By the end of the sixteenth century, the Christian church was splintered. The Roman Catholic church had a reformation of its own, as we saw in Chapter 6, but was now more intransigent than ever. Lutheranism had taken hold, mostly in Scandinavia. Calvin's followers had control over Switzerland, Hungary, and a great deal of influence in France, where the reformers became known as Huguenots. The Anabaptist movement had caught on in Germany, Holland, and in parts of eastern Europe, and continued to be one of the most persecuted religious groups of the times. In Scotland a powerful reformer, John Knox, had started the Presbyterian church, breaking away from the Church of England. The Presbyterian church held to a Calvinist theology, but promoted church government not by a hierarchy of bishops like Roman Catholic and Anglican but by local elders or Presbyters chosen by each congregation. Knox's Presbyterian church became the state church in northern Ireland, while the south stayed true to Rome. The Church of England of course controlled the rest of England, but to the outside it looked remarkably like the Roman Catholic church, with a king instead of a pope. This gave rise to the Puritan movement, which tried to "purify" the Anglican church. Out of the Puritan movement came the Congregational church, also heavily influenced by Calvin, but unlike the Presbyterians they advocated a church government run by the members of the congregation.

With all this turmoil in Western Europe, it is easy to forget about the Orthodox or Eastern church. The Orthodox church was not dormant, but it continued to exert its influence east into Russia. It is important to understand that the Orthodox Christian church teaches that the Body of Christ in fact cannot be divided. Individuals may split from the Church and even call themselves a church, but the Orthodox church remains as the one true Christian church. These schisms we have been examining had relatively little impact on the Orthodox.

To fully appreciate the state of the modern Evangelical church there is one more historical chapter.

End Notes

1. The Anglican Timeline, http://justus.anglican.org/resources/timeline/
   06reformation.html
2. http://tudorhistory.org/primary/supremacy.html, posted by Lara E. Eakins
3. The Anglican Timeline, http://justus.anglican.org/resources/timeline/
   06reformation.html
4. Noll, M. (1997). *Turning Points* (p. 179). Baker Books.
5. English Bible History, http://www.greatsite.com/timeline-english-
   bible-history/william-tyndale.html
6. Foxe's *Book of Martyrs*, Chapter XII.
7. From The Practice of Prelates, http://www.tyndale.org/DeCoursey/practice
   .html.
8. Foxe's *Book of Martyrs*, Chapter XII.
9. Noll, M. (1997). *Turning Points* (p. 197). Baker Books.

11

THE MODERN EVANGELICAL CHURCH

CHAPTER OUTLINE

By the end of the sixteenth century, most of the modern Protestant denominations were established. That does not mean that there were no more changes. For example, the modern Lutheran church is divided into at least two factions: the Evangelical Lutheran Church of America and the more conservative Lutheran Church Missouri Synod. Most of the other denominations that were introduced in the last chapter have also undergone transformations and splits in the last 400 years. However, for most, if not all of them, the unique elements that will impact acoustics and media usage will be very similar within a denominational family. Congregational churches will all be quite similar for our purposes even though there may be theological differences that the members perceive as being quite important.

There are a few important denominations and events that appeared on the scene after 1700 that we must consider. We will be looking primarily at the Baptist church and the Methodist church.

DENOMINATIONS

The word "denomination" comes from the Latin *de nominate*, meaning to name. So in one sense a denomination is simply a grouping of like-minded people calling themselves

Sound of Worship. DOI: 10.1016/B978-0-240-81339-4.00011-9

by a common name. However, in the Christian church the word and indeed the concept of the denomination has been a very important way of identifying parts of the greater Church. It can be used to describe the whole Church. For example, one can refer to the Roman Catholic church and the Greek Orthodox church as being different denominations of the Christian church. However, a more common usage is to refer to the divisions that exist in the Protestant arm of the Church. Denominations typically are formed when members of a particular group become convinced of error in doctrine, practice, or governance and secede, forming a new group, which over time may gather new adherents and evolve into a denomination. Some now question if the denominational differences really matter, suggesting that we are in a post-denominational era. Not surprisingly, conservatives will claim that doctrinal differences are very important, and that denominations do indeed matter and are still relevant.

The Baptist Church

According to one Baptist website, there are over 22 million Baptists in the United States alone. It is almost impossible to define the Baptist church. In fact, there is no Baptist church per se. Rather, the term Baptist refers to a loose and very diverse grouping. Baptist historians do not agree how they came to be. Some of the more conservative Baptists believe that they can trace their lineage back to the very early Church, believing that their form of Christianity represents the only true form of the Church and that they are not true Protestants because they were never a part of the Catholic church, therefore they could not have been part of the Protestant movement. Others see themselves descending from the Anabaptists, still others trace their roots to the Puritan movement in England. There is not even a common theology across Baptists. Some are Calvinist, others lean more toward an Arminian view. There are two elements that seem to define the Baptist; one is an emphasis on water baptism as an extremely important act in the life of a Christian, the other is the centrality of the Word and preaching and practicing correct doctrine, although they may not agree on what correct doctrine is.

The more conservative Baptists such as the Primitive Baptists refuse to allow musical instruments in their worship services, believing like Zwingli that if the Bible does not specifically say that it is permitted, don't do it.

VIEWS ON THE INERRANCY OF SCRIPTURE

Most Christians believe in some level of inerrancy of the scripture. This belief of inerrancy or freedom from error stems from the assertion that the scripture was verbally inspired by God. If the scripture is inspired by God, then the outcome of that process must be without error at least in the original autographs. The Christian view on inspiration is that God "breathed" (inspired from Latin *inspiraer*, to breathe into) the Bible through human agents who wrote it with their own particular styles and vocabularies. The style may be human, but the content is divine. Like many of the Christian doctrines, Biblical inerrancy has been a point of controversy and responsible for a number of rifts and schisms.

Some of these very conservative churches do not belong to any formal denomination but use the moniker Baptist to identify themselves with a particular set of beliefs. As we have noted, however, there does not seem to be a unified Baptist theology. The very conservative believe in the verbal inerrancy of the Bible and will use only the English translation commissioned by King James of England in 1611, known as the King James version. This conservative branch are frequently the ones who claim that they can trace their lineage all the way back to the New Testament figure John the Baptist or Baptizer.

The more progressive or liberal have no problem with modern translations of the Bible, are welcoming of many different theological positions, and use contemporary music in their worship. These are often the Baptists who readily trace their roots to sixteenth-century England. After the founding of the Anglican Church by King Henry the VIII, a movement led by Bishop Hooker and Thomas Cartwright began to call for a *pure* church, one based on Scripture alone. These Puritans were trying to reform the Church from within. Some Puritans believed that reform could not happen in a hierarchical structure like the Anglican church and they called for separation from the Church. Known as the Separatists, they advocated a total break from the Anglicans and also promoted separation between church and state. From these Separatists came the General Baptists. The term General refers to a particular theological position known as General Atonement. During this time there arose a competing form of Baptist known as the Particular Baptists who were not in favor of secession from the

Church but who held somewhat different theological positions from the General Baptists. There is some evidence that some Baptists were very influenced by the Anabaptist movement. If there was Anabaptist influence, the only evidence that can still be seen is the emphasis on Baptism. The pacifist and other political views of the Anabaptist are virtually nonexistent in the Baptist church today.

To summarize then, in the twenty-first century we have tens of thousands of Baptist churches in the United States. Some are ultra conservative and fiercely independent. Others belong to large denominational organizations, some fairly liberal, some conservative. Theologically they are all over the map, and the way worship is conducted will also vary considerably, possibly more than any other grouping of protestant churches. When we see *Baptist* on a marquee all we can know for sure is they advocate immersion baptism, will likely have a baptismal, and that preaching the Word as they interpret it is very important.

The Methodists

By the 1700's, profound changes had swept through society and as we have seen, the Church was not immune to these changes. When Luther was attempting to reform the Catholic church, the notion that there could be more than one expression of a given religion was simply not considered. By the eighteenth century this had all changed. In the so-called new world, it was at least theoretically possible for churches to exist separately, sometimes with radically different ideas and approaches, all with the same protection and civil rights under the state. One group that came into prominence during this explosive time was the Methodist church. Founded by brothers John and Charles Wesley (1703–1791 and 1707–1788, respectively), it was based on ideas that came out of the Pietist movement of the mid-seventeenth century.

Pietism emphasized a personal rather than institutionalized religion, teaching that although sinless perfection was not possible, a life free of conscious sin was not only possible but necessary if one were to follow Christ. John Wesley was a powerful preacher and took his message to the streets. In a time when the Anglican Church had total control over everything religious in England this was an outrage. Preaching was meant to occur during the Mass, if at all, but never in the street outside of the church building! John Wesley believed that the message of Christ was meant for the poor and disenfranchised, not just the "politically correct" of the day. This was a new form of evangelism that the Church had not previously engaged in. He preached a message of "deliverance, recovery and liberty" to a population who desperately needed hope and who had never heard anything like this.

According to John Wesley's journal, he preached 40,000 sermons, often preaching two or three times a day, starting before dawn. Only in his eighties did he stop the pre-dawn sermons. John's brother and partner Charles was a prolific song writer, composing nearly 10,000 hymns. Many of these hymns are found in modern hymn books and are still frequently used in worship in many denominations. The followers of the Wesleys became known as the Methodists because of the methodical way the Wesleys encouraged Bible study and practice. These early Methodists were some of the most successful proponents of the Protestant reform message, and were powerful agents of change as well. By the end of the eighteenth century the battle cry, *Solo Scriptura*, had evolved to mean not the scripture as highest authority, but scripture as the *only* authority. A new branch of Protestants emerged—the Evangelicals.

12

ACOUSTICS OF THE EVANGELICAL STYLE CHURCH

In Chapter 7 we saw that the Celebratory church is dominated by the architecture and the need to create and maintain a holy or sacred space. The Protestant Reformation did away with virtually all the symbolism and the need for a sacred space. Instead of a sacred space designed to be a home for the Mystery of the Eucharist, the Evangelical church is a gathering place for people to hear preaching, a proclamation center. In the Evangelical church the architect, acoustician, and sound system designer will have considerably more latitude in tailoring the acoustics and sound systems to the needs of the people. A cursory examination of the order of a typical evangelical style service outlined in Chapter 8 will reveal that there are actually three types of activities that should be considered in the design of the acoustics and sound systems in Evangelical churches.

Sound of Worship. DOI: 10.1016/B978-0-240-81339-4.00012-0

There is the spoken word from the pulpit to the congregation, there is congregational singing, and there is performance music generally originating on stage and heard by the congregation. Only one of these design goals has a specific objective standard. It is possible, within limits of course, to predict at the design stage the intelligibility or the ability to understand the spoken word in a space with a specific sound system. The other two design goals are not generally specified directly. There are no terms to describe, for example, the "sing-ability" of a room. Unfortunately it is as if there were orthogonal vectors pulling the design in mutually exclusive directions. If the sanctuary is *optimized* for any of the three, the others will suffer. This is what is challenging about the acoustics of the church, especially the Evangelical church. Performance spaces can be optimized for a particular activity. Opera houses can be acoustically optimum for opera, theaters for theater, etc. Evangelical churches are multi-use rooms, and the acoustical and electro-acoustical designs will always be a compromise.

The Acoustics of Preaching

Certainly the most important aspect that needs to be optimized for preaching is the ability to understand, or intelligibility. However, the space must be conducive to preaching as well. Most people have experienced the frustration with trying to speak in the presence of a discrete reflection. Reflections between roughly 40 ms and 0.5 s can make speaking virtually impossible. Sometimes there are surfaces that reflect the talker's unassisted voice back to the pulpit, making it difficult to talk. Most often, however, the talker does not produce enough energy to allow this to happen. If the sound system is not properly designed and installed, the sound from the loudspeakers can reflect off of a surface and arrive at the pulpit, impairing speech.

Intelligibility

Intelligibility is the ability to understand the spoken word. (The technical details of how intelligibility is quantified is fully covered in Chapter 23.) The ability to understand the spoken word is very complex. Usually the assumption is that the talker and listener are native speakers and the focus is on what happens in between the talker and the listener. Before we examine the "space in between," we will briefly examine the role of the talker and listener. The talker clearly plays a role in intelligibility. Variables like enunciation, regional accent, and vocabulary will all affect the ability to understand. The listener is also a big part of the communication loop. If communication is to be effective, that is, for the listener to fully understand what the talker said, the listener has to have adequate hearing, adequate attention span, a shared vocabulary with the talker, and familiarity with any regional accents the talker may have. In addition, the receiver or listener has to have a shared context with the talker. It may be possible for a talker to be talking about a topic totally outside the context of the listener and even though the words are all individually understood, the overall meaning will be lost. This is of great interest to the evangelical. Since the whole point of the church service is to communicate the saving Gospel of Christ to the unsaved who may be in attendance, great emphasis is placed on training pastors in homiletics or the art of preaching.

Clearly the two ends of the communication loop, the talker and the listener, bear some responsibility for the transfer of information. The architect, acoustician, or sound system designer have no control over the ends of the communication loop. However, the message is transmitted though the air in the space. If that air is bounded by surfaces (i.e., not outside), by definition there is acoustics. The acoustical behavior of that bounded space is the result of the architecture. The architect, wittingly or not, is creating acoustic space. So the question is, how does the acoustic space affect intelligibility, if at all? Even though a great deal of the responsibility for intelligibility falls to the sound system designer, there are four factors that the architect and acoustician need to consider with respect to intelligibility: noise, reverberation, reflecting/focusing surfaces, and what might be called sound-system–friendly architecture.

Noise

Noise (covered more fully in Chapter 22) is unwanted sound. The architect/mechanical/acoustical team need to keep the noise levels in the evangelical church below a PNC 25, not because of the sense of awe associated with quiet space but because noise is one of the strong contributors to the degradation of intelligibility. All the potential sources of noise must be considered: HVAC, traffic, public transportation systems, flight paths, emergency vehicles, and any other source of noise that may interfere with the presentation of the Gospel.

Reverberation

Reverberation (covered more fully in Chapter 21) is how long sound bounces around in the space. In terms of how reverberation effects intelligibility, it works in much the same way as noise, with one important difference. Noise is uncorrelated to the speech whereas reverberation is correlated as each reflection is in essence a copy, albeit spectrally shaped, of the original signal. Correlated signals may actually be more effective in masking the direct sound and therefore be more deleterious than random noise. In Evangelical churches without pipe organs there is no reason to have reverberant space. It shouldn't be subjectively "dead," but it should have a very short reverberation time. If there is a pipe organ planned it should be understood that the requirements for a pipe organ are in direct conflict with the requirements for intelligibility. It does not mean that if a church installs a pipe organ there will be no intelligibility. However, if the room is optimized for the organ and has a mid-band reverberation time in the 2.5 to 3 second range, the sound system will have to be more carefully designed, and should consist of loudspeakers with high directivity if intelligibility goals are to be met.

Reflecting/Focusing Surfaces

When the architecture includes surfaces that reflect specular energy back to listeners such that energy reaches a listener with more than 25 milliseconds delay, the reflection or "slap" may be heard as a separate event, or echo of the original signal. When this occurs intelligibility really suffers. Shorter latency reflections are not experienced as discrete events, but their effects on the spectral content of the material can be devastating to intelligibility as well.

The back wall is often the biggest culprit. The back wall can reflect sound back to the pulpit, making it difficult to preach and/ or it can reflect sound to select areas of the congregation rendering some areas virtually unintelligible. Curved surfaces, especially concave surfaces with respect to the seating area, are especially bad and should be avoided at all costs. Convex surfaces are more acoustics-friendly, but can still cause intelligibility problems in some areas. The worst-case condition is when there is a concave back wall and the center of that segment of a circle is the pulpit! All the sound energy is then focused right back at the pulpit, making it virtually impossible to speak. If curved surfaces must be used the church needs to plan on installing a significant amount of broadband absorption on the curved surfaces so that focused reflections are minimized or eliminated.

Of course, flat surfaces also reflect sound and should be evaluated for potentially problematic reflections.

Sound-System–Friendly Architecture

Architects need to understand that sound systems, especially loudspeakers, need to be installed in specific places for important reasons. There are numerous types of loudspeaker systems and approaches to implementing sound reinforcement systems that are covered in Chapter 20. As a general rule, in order to insure that there is intelligibility in every seat in the congregation, everyone seated in the congregation must have an unobstructed view of a loudspeaker. Often the optimal location for mounting loudspeaker systems will be directly above the pulpit. Especially in a larger Evangelical church this area needs to be left clear of visual symbols or elements that would get in the way of a loudspeaker cluster. In addition, structural support should be provided, especially in larger rooms to accommodate the loudspeaker system. Of course, if the planning of the facility involves the architect and the acoustical consultant as well as the media systems contractor, the audio and video systems can be integrated into the architecture and not simply added as an afterthought.

If the noise is kept to a minimum, the reverberation is controlled and kept appropriately low, surfaces that are creating harmful reflections are appropriately treated, focusing surfaces are avoided, and there is system-friendly architecture, then the sound system contractor can install a sound system that can be guaranteed to provide intelligible sound to every seat in the room.

Acoustics for Congregational Singing

If speech intelligibility is the most important design goal, congregation singing is a close second. In this tradition, when the congregation joins in song, it is important for them to be able to hear themselves! All surfaces of the space contribute to some degree to letting the congregation hear themselves. The most important surface, however, is the ceiling. The biggest acoustical mistake that an Evangelical church can make is to install an absorptive ceiling. It is sad to see churches with acoustical tile ceilings. What often happens is the poor song leader is up there trying desperately to get the congregation singing and often they in fact are singing to the best of their ability, but to the song leader their attempt at praise seems lifeless and feeble. And, since they cannot hear themselves, they give up trying.

To optimize the acoustics for congregational singing, the ceiling should be between 9 and 14 feet above average standing height. A ceiling at this height will provide reflections back to the congregation in the range of 16 to 24 milliseconds. Reflections in this range are generally not heard as discrete events but will impart a sense of loudness to the singing. Ceilings that are lower than this will create reflections that are too early and will be most likely perceived as irritating. Ceilings that are higher or acoustically further away produce reflections that arrive so late they will be heard as discrete echoes or slaps, which is also not desirable.

Figure 12.1 The Use of Reflecting "Clouds". 12 Stone Church.

It is best if the ceiling is diffusive and not simply flat. Diffusion will help impart a sense of envelopment to those singing, enhancing the worship experience. If it is not possible to place a ceiling such that it provides reflections in this range, acoustic reflectors sometimes known as clouds can be suspended from the deck down to the desired heights as seen in Figure 12.1. Again, the clouds should be diffusive and not simply specular.

It is important to note that providing relatively early reflections back to singers is not the same as creating a reverberant space. A live ceiling will likely play a role in the creation of a reverberant space, but in the evangelical church, unless there is a pipe organ, as noted earlier, there is no real need for a true reverberant field. The reverberation time can be controlled with strategically placed absorbers on the back and side walls as well as utilizing padded pews and pew backs. If there is a pipe organ planned for the space, it is a better strategy to make the ceiling higher to allow for some volume, thereby creating a larger acoustic space with some mid-band reverberation and suspend diffusive clouds to an appropriate height above the congregation.

Acoustics for Performance

The third design goal is the acoustics for performance. Evangelical churches will likely have a choir and probably use piano and organ (either pipe or electronic). Of course, many use contemporary instruments like guitars and drums. Although this is not the most important element in the evangelical worship style, there is an expectation that the choir will sound good and project well into the congregation. This requires reflecting surfaces above the choir area angled to project the sound of the choir into the space. The height is not as critical as the clouds for the congregational singing, but 7 to 15 feet is a good range. Portable acoustical shells like the one in Figure 12.2 may not fit into the decor of the church, but reflecting surfaces similar in concept can often be integrated into the architecture.

Figure 12.2 Typical Choir Shell. (Courtesy Wenger Corp.)

Earlier we mentioned the necessity for controlling the reflections off of the back wall so that the preacher will not struggle with discrete late reflections, which make it difficult to talk. The same is true for the choir and instrumentalists. The difference is that musicians would rather play and sing into a live and even slightly reverberant space than into a dead space. Like the congregation, the choir needs to hear itself. This makes the design of the acoustics of the choir space a bit more complex as we want the sound to both come back to the choir and be projected out. Architects frequently figure that the sound system contractors will electronically enhance the choir and also provide them with the foldback or monitors that they need to hear themselves. Whereas it is true that it is possible to reinforce the choir and even provide them with monitors, it is better to accomplish this acoustically whenever possible.

If the design is optimal for any one of these goals, the other two will suffer to some degree. It is the responsibility of the design team to make sure they understand the expectations of the church so that the appropriate compromises can be made. However, the predominate goal is intelligibility. Hearing and understanding the Word trumps the music!

We will examine a few select Evangelical churches to show how various designers have dealt with the challenges of the evangelical worship style.

Case Study

Parkview Community Church.

Parkview Community Church, Glen Ellyn, Illinois

Architects	Bruce Koenigsberg and Mark Elmore Associates
Acoustical Consultant	EASI Evanston, IL
Sound System Design	EASI Evanston, IL
System Installer	Parkview Church and Technotrix, Bourbonais, IL

Acoustic Specification (measured)

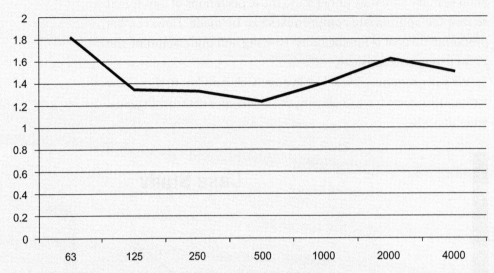

Figure A Parkview Community Church Reverberation Time, Empty.

F (Hz)	T30 (s)	T20 (s)	T10 (s)	EDT (s)	C80 (dB)	C50 (dB)	D50 (%)	BR
63	2.039	2.194	1.765	0.918	8.67	8.39	87.36	0.998
125	1.409	1.332	1.951	1.075	10.03	9.73	90.39	
250	1.17	1.165	0.279	0.036	15.67	15.24	97.1	
500	1.237	1.313	1.44	0.037	13.83	12.68	94.88	
1000	1.247	1.19	1.209	1.373	9.7	7.74	85.61	
2000	1.393	1.293	1.042	1.872	11.03	9	88.81	
4000	1.25	1.233	1.188	1.304	9.77	6.73	82.5	
8000	0.872	0.787	0.938	0.456	14.82	11.55	93.46	

Table A Parkview Community Church Acoustical Parameters (measured per ISO3382).

Design Narrative

The new building sanctuary was designed for 550 people. We wanted to try to keep the feeling of closeness or intimacy of the older space while seating more worshippers. With higher ceilings and a much wider room, we knew this would be a challenge. In addition to maintaining closeness, we wanted to avoid long reverberation times, which would muddy the message. We also had the benefit of starting from scratch with a brand new audio system design, which permitted us to rethink our music ministry.

We began working with our acoustical consultant early in the sanctuary design phase. We held a number of meetings with the acoustical consultant, the architect, the builder, and the sound team. This enabled us to design the room right from the start and not have to fix it later once the gear was installed. There were a number of acoustical concerns. First and foremost, we wanted the sanctuary to provide the support to congregational singing, without being excessively reverberant. We were also concerned about sound bleed-through from adjacent rooms and hallways. To help this the acoustical consultant specified resilient channel mounting for the plasterboard and doubling up the plasterboard thickness on certain walls. Sound bat insulation was added between all walls of the sanctuary. We were also concerned about HVAC noise and the acoustical consultant helped to specify the type of ductwork, isolation joints, and the overall NC ratings for the HVAC system.

Per the acoustical specification, a series of acoustical panels with custom fabric coverings were installed on the rear wall of the room. A sprayed-on insulation foam in the nonwindow areas of the skylight was used to reduce reflections from the skylight area. A saw-tooth side wall design was designed for the side walls to provide useful reflections back to the congregation. A cultured stone wall was used to provide some sound diffusion on the stage. The stage was poured concrete covered with carpeting. Stage pocket boxes were installed across the stage for audio wiring. The main sanctuary floor is sloped toward the stage and is covered with carpet. The chairs have padded foam with cloth covering. The ceiling around the sides of the skylights was designed by our architect and acoustical consultant as a series of hanging clouds of dropped ceiling grids. These clouds reflect some of the audience singing back to themselves.

The acoustical consultant also specified the type of loudspeaker cluster to use (three Martin WT3s in bi-amp mode) as well as their placement. For side-fill loudspeaker we used Martin W2s. Parkview elected to install the sound system ourselves, but we retained Technotrix to hang the loudspeaker clusters. When we installed the system, the consultant assisted with the aiming of the loudspeakers and tuning of the room with our DSP system.

To help combat stage audio levels we opted to go with an Aviom system with 13 personal stage mixers. Besides the drum kit, piano, horns, and violin, all other instruments use direct boxes into the mixer.

We are very pleased with the outcome of the acoustics and the sound system of this new space. The room has a nice natural resonance sound and does not feel too acoustically dead at all. We have been worshipping in it for over three years now and have not had to make any substantial modifications. Starting with an acoustical consultant at the onset is best way to go—particularly for new building designs.

(Submitted by Mark Ahlenius, Parkview Community Church)

Case Study

Anonymous Christian Church

The following is excerpted from a report on a church with numerous problems. First, the letter sent to the pastor.

Ref: Acoustical Analysis of Sanctuary

Dear Pastor,

It is my pleasure to present the results of the acoustical analysis of the main sanctuary at your facility. This report is based on measurements conducted in the church in December 2005, subsequent analysis of the data gathered, and acoustical modeling using a computer-aided virtual-room design of the space. Recommendations for improving the acoustical environment are provided, with suggested material and mounting locations. It is our hope that by implementing these recommendations, significant improvements will be realized, resulting in a worship space more conducive to contemporary services.

As will be seen, there are three main issues to consider. The first is the room's geometry. Shapes such as circles or octagons are extremely difficult spaces, typically characterized by poor acoustics. Next are the room's surface treatments, which can lead to an excessively reverberant environment. Finally is the issue of noise, primarily that typically generated by the HVAC system and other mechanical systems. Failure to deal with all three of these concerns will result in less than adequate results.

We hope that the information provided in this report will assist the church board in understanding the problems faced and encourage the establishment of a realistic budget to finance the changes recommended herein.

Please feel free to contact me with any questions you may have. I'll be more than happy to assist in any way possible.

Here, in part, is the body of the report.

Analysis and Recommendations

Architectural geometry plays a critical role in the acoustical characteristics of any room. Shapes that support good acoustics are usually asymmetrical or include significant variations in wall, ceiling, and seating shapes. Fan-shaped rooms or "shoe box" geometries can be designed to encourage supportive rather than destructive reflections and/or excessive reverberation.

On the other hand, shapes that exhibit geometric symmetry, such as circular- or dome-shaped rooms, are very difficult. Octagonal-shaped rooms are essentially round rooms with flattened walls. This presents an extremely difficult situation in that the eight sides form four sets of flat, opposing

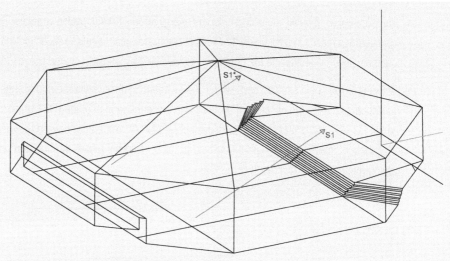

Figure 1 Wire Frame Model of the Sanctuary.

surfaces that support what are referred to as "flutter echoes." We can develop a good feel for this phenomenon by listening to the audio file entitled "East wall to West wall—TOA speaker," which accompanies this report. This .wav file audibly demonstrates acoustical energy bouncing back and forth in the sanctuary. These flutter echoes can be heard as the energy decays over time.

Figure 2 provides a visual representation of this event in the form of an Energy–Time Curve. The vertical axis represents amplitude or loudness, and the horizontal axis shows time in milliseconds. A total time span of 773.68 milliseconds, or about 3/4 of a second, is shown.

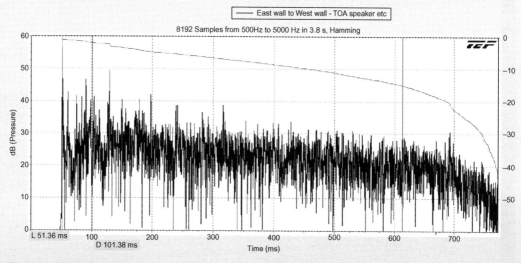

Figure 2 Energy Time Curve.

The evenly spaced peaks in the chart identify energy returns from opposing surfaces. This geometric configuration results in a condition that causes percussive instruments to sound cluttered and harsh. The additional energy blends into the original sound and destroys musical clarity. The result is a din of sound.

In Figure 3, we see a similar chart, which shows energy emitted from the house sound system loudspeakers. Note the absence of periodic energy peaks seen and heard in Figure 2. This is due to the fact that the front of house loudspeakers are aimed at the floor toward the rear of the room. The angle of incidence is about 25 degrees, so the reflection bounces into the rear wall with some absorption occurring as a result.

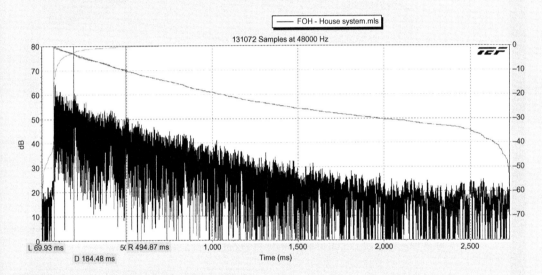

Figure 3

The solution is to install acoustical treatment to as many acoustically hard surfaces as possible. Using a three-dimensional room created in EASE, a computer-aided modeling program, we have studied the room's acoustics and arrived at some recommendations. Figure 4 (not shown here) tabulates reverb times calculated by EASE. Table 1 (not shown here) shows octave-band data gathered during the site survey. We would hope to lower these times substantially, in a range between 0.7 and 1.0 second.

Using EASE to observe the changes we might expect with the addition of acoustical treatment, we find that substantial reductions in RT_{60} can be realized. Figure 5 (not shown here) shows the reduction realized by covering the walls and ceiling. Products manufactured by TECTUM, Inc. were used in the model to achieve the results displayed. Details on the materials recommended are attached for your review.

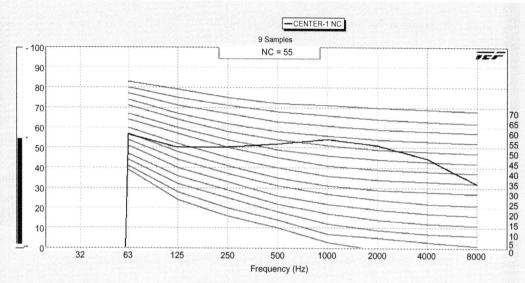

Figure 6 NC as Found.

In addition to lowering RT_{60}, the acoustical treatment will help lower the levels of late arrivals, which will make the room a much more pleasant environment in which to worship.

Finally, we come to the subject of ambient noise. I must say I was shocked when I heard the air conditioning system turn on for the first time. It is, without a doubt, one of the noisiest HVAC systems I've ever heard! Noise levels generated by this system are unacceptable for virtually any type of room, including gymnasiums. In my opinion, high ambient noise levels are a more serious problem than high reverberation times.

Design criteria used to engineer and install heating, ventilation and air conditioning (HVAC) systems can be found in the attached document titled, Ranges of Indoor Design Goals for HVAC System Noise Control, published by ASHRAE, the American Society of Heating, Refrigeration, and Air-Conditioning Engineers. By referring to the appropriate room type, churches and schools, we see that acceptable range of noise levels for sanctuaries are 25–35 A-weighted SPL or NC 25–35. Data measured during the site survey is shown in Figures 6 and 7. When compared with the recommended allowable noise levels established by ASHRAE, 25–35 dBA/NC 20–30, we can easily see that the system is far too noisy. Without correcting this situation, the room will never be able to fulfill its roll as a House of Worship. Indeed, if we were considering performance criteria for a gymnasium, the design would call for 40–50 dBA/NC 35–45!

In conclusion, the three issues just explored are:

1. Room geometry—we are stuck with this!
2. Resulting acoustical characteristics—treatable
3. High ambient noise levels—a tough one

Obviously, the room's geometry cannot be changed. We can affect the room's acoustical nature by covering ceiling and wall surfaces with appropriate acoustical treatment. In a small way, we will be changing not only the room surfaces but also slightly changing the geometry by angling the acoustical materials during installation. Finally, the HVAC system remains a very difficult problem and will require the expertise of an expert in test and balance of such equipment. In my estimation, the method of air distribution is totally wrong and lowering noise levels appropriately may require substantial modification of the system if not a complete redesign of the duct system. This will not be easy or inexpensive; however, as stated earlier, it is critical to making the room work.

Cost estimates for the acoustical product needed will be forthcoming as soon as the manufacturer has completed their analysis of the information we have already provided them. Costs for the installation of these products will need to be provided by a local contractor. Costs for correcting the HVAC system will not be available until a certified test and balance firm has provided their analysis of possible fixes.

I hope the information presented in this report will be helpful to the church in making decisions on how to proceed with remedial efforts to correct the problems discussed. Please feel free to contact me with any questions that may arise after you've had a chance to digest the report. I'll be happy to assist in any way possible and look forward to working with you as the project develops.

We have included this case study to illustrate the sorts of problems that can occur when some basic steps are not followed. In this case the church chose to use a church design/build firm who "has built many of these buildings successfully." There certainly are some good design/build firms building fine churches. However, the buyer needs to be aware of these sorts of shapes, and especially designs that do not specify any sort of noise criteria.

3

EXPERIENTIAL WORSHIP

THE EXPERIENTIAL STYLE

CHAPTER OUTLINE

In a sense, all Christian worship is experiential. In the Celebratory church, the believers experience the grace of God via the Sacrament of Eucharist. In the Evangelical church believers hear the word of God and experience salvation or find a sense of direction for living. For some churches, however, it is the *experience* that defines the worship. Even though we might be hard pressed to find a church that claims to worship in the experiential style, it is clear that there are a significant number of churches that conduct worship in a manner that emphasizes the experience. In these churches the worship service can only be described as a performance, and they don't seem to be troubled by this at all. This stands in stark contrast to the celebratory style, especially the Roman Catholic version where referring to the Mass as a performance would be seen as a grave insult. When there is an emphasis on experience, worship takes on a unique quality, which has important ramifications for acoustics and media systems. The experiential church is likely to have sophisticated media systems that are a significant part of the experience.

There are broadly two forms of the experiential style of worship: the seeker-sensitive or seeker-friendly church and those who trace their history to the Pentecostal movement. The seeker-friendly church is a relatively modern form of church. Many credit Dr. Robert Schuller of the Crystal Cathedral of Orange

Sound of Worship. DOI: 10.1016/B978-0-240-81339-4.00013-2

County, California, as being the founder of the seeker-sensitive or seeker-friendly movement. Schuller had a vision to build a church for people who would never consider darkening the door of a conventional church. He started out renting space from a local drive-in movie theater. His message of "possibility thinking" is based largely on the teachings of Norman Vincent Peale, who was known for his message outlined in his popular book, *The Power of Positive Thinking*. Schuller certainly did succeed in building a church that was welcoming to everybody, but his critics wonder exactly what people are being welcomed into. Is this really the Church, the body of Christ? Or is it a feel-good "gospel" that skirts the "heavy" issues of sin and evil? Questions have plagued the seeker-friendly movement for many years. In 1998 *Evangelicals Now* published an article that reviewed the Seeker-Friendly Church (SFC) movement.

> *Should the Sunday service focus on the seekers/unchurched or on believers? The seeker-focused service (as made popular by Willow Creek) has the following elements: multiple communication forms (commonly music, drama, testimony, teaching, questions-and-answers); excellent production/ artistic quality; low or no audience participation (but it does challenge the onlooker into participation with Christianity); the teaching/talk is purely apologetic and evangelistic.*
>
> *One fundamental of the SFC movement is that the same service cannot target both seekers and believers. It is said that you can reach one or the other but not both. Most thoughtful proponents of the straight seeker-service agree that it is not worship, and cannot give Christians proper nurture. Therefore, in many churches in the SFC movement, a "seeker-focused" service is targeted for the unchurched, and some other weekly believer-focused service becomes the worship service for believers. In many other SFC churches, there is only one main weekly service which serves believers and the unchurched. This is called "seeker-sensitive" worship.*[1]

Recently, more evangelical forms of the seeker-friendly church have emerged. Willow Creek Community Church in South Barrington, IL, founded in early 1975 by Bill Hybels, is now a mega church with more of a mainline Christian message. They also go to great lengths to create an experience on Sunday that is not intimidating or at all uncomfortable to someone who may be seeking for God. Their strategy seems to be to get people in the door on Sunday by whatever means possible, but it doesn't end there. There are numerous smaller, more intimate settings where a new believer who may have been initially drawn to the experience can be mentored and grow in the faith. Another example of a very successful seeker-sensitive mega church with an evangelical message is Saddleback Valley Community Church in Southern California, led by Pastor Rick Warren. Warren is the author of the bestseller series of books on the Purpose Driven Life. Like Willow Creek, Saddleback may make the Sunday worship seeker-sensitive, but they are committed to growing committed, mature Christians.

A worshipper who desired to remain anonymous shared this first-person account of a visit to a large seeker-friendly church in a major metropolitan area.

I arrived a bit late for the second Sunday morning service. I was pleasantly surprised as I drove into the "campus" to see there were parking attendants who directed me to a parking area near the main entrance reserved for late comers. It occurred to me that although it was very convenient and even thoughtful, this sort of psychology might back-fire! The campus was beautiful, immaculately landscaped and serene. The building was unusual in that it was difficult to get a sense of the scale of the structure. It was hard to tell if this was a building that would seat 500 or 5000. Every detail said welcome. As soon as I entered the building I was directed to the "in-house-famous-name-coffee" station for a free hot cup of coffee. Armed with my coffee, I entered the tunnel leading into the worship area. It reminded me of the entrance to a sports facility. You could not see the sanctuary until you got to the exit of the tunnel. The room was vast! There was tiered seating with a fairly steep rake, all focusing on a large stage surrounded by three very bright hi-def projection screens. I was met by a greeter who asked how many in my party. I told him I was alone and he directed me to a seat half-way up the section. I admit to being a bit self conscious as I really don't much enjoy visiting strange churches. I would have preferred to enter in the back of a church and sneak into the last pew unnoticed. I started up the steps to my assigned seat, tripped on the stairs almost spilling my coffee, my preference for a discrete entrance blown! I took my seat and was greeted warmly by those on either side.

In a few moments, the Vari-lites® swooped down and focused on the drummer seated in a closed-off drum booth. As he started playing, I could feel the kick drum in my chest. He was quickly joined on stage by the rest of the band and they launched into a contemporary worship song. It was loud, but not deafening. I knew the song and wanted to sing along, but there was no way I could ever hear myself sing much less anyone else around me. The next song was unfamiliar, but more contemplative and melodic, but it brought the energy way down. As this song drew to a close, the lead guitar player started praying for God's presence to be in this room. The band vamped underneath the prayer which ended with a high power drum fill and a new song started with more energy than before. I remember thinking to myself, These guys are good! I got the sense that I was being manipulated, but I didn't care, it was fun!

The singing went on for about 10 more minutes, after which there was a brief time of announcements about the life of the church and other meetings and services that people could get involved with. Then the preacher for the day came on stage with a high stool and sat down to deliver the message of the morning. The hi-def screens immediately showed the scripture text. As the talk unfolded, the screens

*followed along with maps, graphics, and photographs all
supporting what was being said. The preaching lasted
20 minutes max. I can't remember ever hearing preaching like
this. There was nothing said that was challenging in the way
that Evangelical preaching often is. "Sinners in the hands of
an angry God" it most certainly was not! After the preaching
they announced that there would be an offering taken and
the worship band came back on and played some more music
as baskets were passed around. I decided that this was my cue
to leave as I wanted to beat the rush of traffic as the church let
out. So, as everyone stood to pray I quietly left.*[2]

The Pentecostal Church

Others who utilize an experiential form of worship are the
churches who trace their roots to the Pentecostal Movement. The
Pentecostals trace their roots back to the Holiness Movement
of the late nineteenth century. An important part of Pentecostal
theology and praxis is the experience of the baptism in the Holy
Spirit. This term first coined by John Fletcher, a colleague of John
Wesley of Methodist fame in the mid eighteenth century, refers to
"an experience which brought spiritual power to the recipient as
well as inner cleansing."[3] This baptism was eventually connected
with the practice of speaking in tongues or *glossolalia.*

GLOSSOLALIA

Glossolalia, or speaking in tongues, refers to the practice
of speaking in a language that the speaker has never learned.
Christians who believe in glossolalia teach that there is a baptism in the Holy Spirit that may or may not occur simultane-
ously with water baptism. This additional baptism is what
was experienced by the disciples of Jesus after his ascension.
According to the Biblical account recorded in the book of Acts
Chapter 2, when God sent the Holy Spirit to descend upon
the disciples, they began to speak in languages they had not

learned. Technically, the Biblical account is an example of *xenoglossia*, or speaking an actual language the speaker has not learned. Some who practice glossolalia assert that they are speaking real human language (xenoglossia), however the majority say that they are speaking an ecstatic worship language rather than a real human language. Sometimes there will be an interpretation of the unknown language for the benefit of those present who may not understand. The proponents of this practice are generally the Pentecostals and a newer, off-shoot of the Pentecostals known as the Charismatics.

The early roots of the Pentecostal Church in the early 1900s was both a religious and cultural movement. Most historians point to the Azusa Street Church in Los Angeles as the true beginnings of the Pentecostal experience.

The Azusa Street movement seems to have been a merger of white American holiness religion with worship styles derived from the African-American Christian tradition which had developed since the days of chattel slavery in the South. The expressive worship and praise at Azusa Street Church, which included shouting and dancing, had been common among Appalachian whites as well as Southern blacks. The admixture of tongues and other charisms with black music and worship styles created a new and indigenous form of Pentecostalism that was to prove extremely attractive to disinherited and deprived people, both in America and other nations of the world.

The interracial aspects of the movement in Los Angeles were a striking exception to the racism and segregation of the times. The phenomenon of blacks and whites worshipping together under a black pastor seemed incredible to many observers. The ethos of the meeting was captured by Frank Bartleman, a white Azusa participant, when he said of Azusa Street, "The color line was washed away in the blood." Indeed, people from all the ethnic minorities of

Los Angeles, a city which Bartleman called "the American Jerusalem," were represented at Azusa Street.[4]

Worship as practiced by those with Pentecostal origins, which according to some are second in number only to the Roman Catholic church,[5] can appear to the outsider as controlled chaos! It is very participatory and involves loud music and energetic preaching interrupted with ecstatic outbursts from the congregation. The worship will often involve performed music and, especially in African-American churches, gospel choir music. The underlying element is the *experience* of the outpouring of the Holy Spirit and the power of God to change lives. This experience is sought by those who attend and everything—the singing, the performing, the preaching, the healing, the speaking in tongues—all combine to provide the experience.

To put the Pentecostal Movement into perspective, we will briefly examine the history and roots of the denomination in Chapter 14.

End Notes

1. Evangelistic Worship: The problems of praise and worship and making it relevant to all—And a review of the SFC (Seeker-Friendly Church Movement), *Evangelicals Now*, December 1998.
2. Personal Correspondence 9/2009.
3. Synan, V. The Origins of The Pentecostal Movement, www.oru.edu/university/library/holyspirit/pentorg1.html, accessed 3/10.
4. Ibid.
5. Ibid.

14

THE PENTECOSTALS AND THE CHARISMATIC MOVEMENT

CHAPTER OUTLINE

The Pentecostals

PENTECOSTAL

The word *Pentecost* refers to an event related in the New Testament: on the Jewish holiday known as Pentecost, which occurs 50 days after Passover, the Holy Spirit descended on the Disciples of Jesus. This event took place after the resurrected Jesus had ascended into heaven, leaving his followers confused and afraid. When Pentecost occurred, the Disciples were "filled with the Holy Spirit and began to speak in other tongues as the Spirit enabled them." This experience turned the timid fearful Disciples into bold outspoken Evangelists who took the message of Christ to the whole known world. In the broadest of all possible terms, the Pentecostal Movement seeks to duplicate that experience in all Christians.

Many historians of the Pentecostal movement trace its roots back to John Wesley. In addition to promoting individual Bible study and personal piety, Wesley taught that conversion should include some sort of dramatic experience. It was this inclusion of an experience that became the hallmark of the Pentecostals.

Sound of Worship. DOI: 10.1016/B978-0-240-81339-4.00014-4

In the years following Wesley, the Methodists, especially in the New World, became more and more focused on the experience of what they called sanctification. These fiery preachers would hold meetings where all sorts of unusual phenomena were observed. In the early 1800s, revival meetings broke out in various parts of the young country, most notably in Cane-Ridge, Kentucky. This was the birthplace of the camp revival meeting. By 1801 the revival meetings in Cane-Ridge were drawing crowds of anywhere between 10,000 and 25,000 people at a time. All sorts of "Godly Hysteria" broke out with people jerking uncontrollably, barking like dogs, and speaking in unknown languages.

Within a decade, the Western revival, which began in Cane-Ridge in 1800, became more institutionalized as the camp meeting became a regular part of American religious life. By 1830 the more frenzied aspects of the revival had become little more than a memory, while the primary concern shifted from experience to doctrine. However the emotional type of religion continued to exist, especially in the frontier and in the South.

This new form of Methodist faith became known as the Holiness movement, and it established footholds throughout the country. Roughly 100 years later there was a revival at Azusa Street in Los Angeles. The individual at the center of this revival was W.J. Seymour, a poorly educated African–American holiness preacher. Seymour had traveled to Houston to study at a Holiness Bible school run by a man named Parham. The cultural rules of the time made it impossible for Seymour to formally study at the school, but Parham took him in anyway, permitting him to attend classes during the daytime. Word of this determined young preacher got around and in 1906 he accepted an invitation to pastor a small Holiness church in Los Angeles. On his way to take this job, he stopped at various Holiness centers or churches to preach. On April 9, 1906, Seymour was preaching in a home on Bonnie Brae Street in Los Angeles. "Seymour and seven others fell to the floor in religious ecstasy, speaking with other tongues."

Word of these unusual events traveled quickly and people crowed to the Bonnie Brae Street home to hear Seymour.

The home was far too small to accommodate the crowds so they searched the town for a better place to meet and found an abandoned Methodist church on Azusa Street. As soon as Seymour started preaching at Azusa Street, revival hit! This was such a large scale and dramatic revival that it caught the eye of first the local, then the world press. All sorts of unusual behaviors were reported—speaking in unknown tongues, healings—in short the whole gamut of Holiness movement experience. For most historians, the Azusa Street revival marks the beginning of the modern Pentecostal movement in the United States. Generally, the denominations that embrace Holiness theology and consider themselves Pentecostal trace their roots back to this event. "In historical perspective, the Pentecostal movement was the child of the Holiness movement which in turn was the child of Methodism. Practically all the early Pentecostal leaders were firm advocates of sanctification as a second work of grace and simply added the 'Pentecostal baptism' with the evidence of speaking in tongues as a third blessing superimposed on the other two."

By the 1960s, the Pentecostal church had become so mainstream that the famous Evangelist Billy Graham officiated at the dedication of Oral Roberts University in Tulsa, Oklahoma (ORU). ORU was the first Pentecostal university and the first Pentecostal seminary to offer advanced degrees. However, in 1968 Roberts did an interesting thing. Apparently feeling that he needed even broader support, he joined the Methodist church. In his early years Roberts had enjoyed strong support for his faith healing ministry from the Pentecostal community. Later he found that he was gaining acceptance and financial support from a sort of Neo-Pentecostal.

The Charismatic Movement

These were Christians who did not necessarily buy into the traditional theology of the Pentecostal church, nor were they even attending Pentecostal churches. They however sought after similar experiences of being Spirit Filled, and many of them spoke in tongues sometimes to the dismay of the more traditional, non-Pentecostal churches to which they belonged. This movement gained in prominence throughout the 1960s. So when Roberts left the Pentecostal church, joining the Methodist church all the while promising never to leave his Pentecostal theological roots, he was actually tapping into the beginnings of possibly the most important religious movement of the late twentieth century. These Neo-Pentecostals referred to themselves as the Charismatic Movement to distinguish themselves from the mainstream Pentecostals. By the end of the 1960s there was such momentum that Charismatics were popping up in virtually every Protestant denomination and even in some Roman Catholic and Orthodox churches.

Not surprisingly there was some push-back to the Charismatic movement. Some organized churches like the Episcopal church denounced the Charismatic movement. Others like the Orthodox, though not condemning charismatic practice, questioned the need for such an experience.

A very dedicated and sincere clergyman, as best we could gather from speaking with him, visited our monastery during the Paschal Season. He entered the monastery Church with one of his followers, who had "converted" to Orthodoxy under the influence of this Priest, who was active in the charismatic movement among Orthodox, from a Protestant Pentecostal sect. Both the Priest and his companion were somewhat uneasy on entering the Church, not knowing what to do in confronting a place of worship free of pews (as all Orthodox Churches should be), where men and women were standing on opposite sides.

The two visitors began to follow the service in one of the standard Lenten service books used in the more modernized Orthodox jurisdictions or, of necessity, by English-speaking Orthodox. It was immediately apparent that they were lost, as they discovered that their books contained only portions of the prescribed services—something tremendously unfortunate, since the Lenten services are so magnificent in their structure. Added to the fact that they expected the service to last ninety minutes, when in fact it was almost three hours in length, the inadequacy of their service books obviously frustrated them, since they began to pray aloud and to make loud exclamations—"Praise the Lord... Amen."

Finally, one of the Faithful, not knowing that the Priest, who had no beard and wore western clerical garb, was Orthodox, quietly informed him that Orthodox worship in silence and solemnity, with great attention to inner prayer. He explained that the clergyman and his companion were interrupting the service and thus cutting the Faithful off from the mystical quiet that is so conducive to true worship. The Priest and his friend stood through the rest of the service quietly and apparently moved by the worshipful atmosphere.

In Churches where Priests look and act like lay people, where quiet meditation and spiritual chanting have been replaced by organs and the theatre, where pews dull our senses and cater to our bodies, where physical preparation for an encounter with the divine (fasting, prostrations, etc.) is inadequate—is it not exactly here that we find Pentecostal emotionalism spreading like fire among the simple Faithful and the unfulfilled believers? We need not even answer the question.[1]

Other churches like the Presbyterians made speaking in tongues an accepted practice. Even though the Charismatic movement did not have universal support among Christians, it is clear that the Charismatics had a large impact on Christian worship. They brought an emotional and "heart" component to churches that were either intellectual and "heady" or steeped in mystical and symbolic tradition. Nowhere was this more apparent than in the music of the Christian church. The most prolific songwriters of the late twentieth century came out of the Charismatic movement. This new "praise music" was more about experience and emotion and less about theology. The music embraced the new forms of expression that were so powerful in the 1960s and 1970s.

In the early 1970s the term "Christian Rock" was an oxymoron to much of the Christian church. And, although it still is taboo for some of the most conservative, Contemporary Christian Music is now a force to be reckoned with, not just in the Church but commercially as well. In terms of acoustics and sound system usage, nothing in the history of the Church has had as much direct influence. The vast majority of Protestant churches in the United States use contemporary music in their services. Even churches that are very conservative—churches who would frown on speaking in tongues or conducting faith healing services—will use music that came directly out of the Charismatic movement. Sometimes these churches are small and require a modest sound system. Others are gigantic mega-churches requiring systems costing in the hundreds of thousands of dollars. For each of these churches the acoustics and media systems now play a far greater role then ever before. And it is not just Protestant churches. The Roman Catholic church faces even greater challenges with respect to contemporary music. Many Catholic churches have different Masses to serve different parts of the community. Some will have contemporary Masses using modern music and amplified instruments. In urban centers this is often attempted in traditional cathedral-type buildings with reverberation times in excess of four seconds. The acoustics of these spaces is optimal for pipe organ and chant, but decidedly not optimal for experiential worship!

African–American Churches

Another very important development in the last 150 or so years is the emergence of an organized African–American church. There has always been an African–American church stretching back to the dark days of slavery. The African–American theologian and historian Dr. James Cone writes:

When blacks investigated their religious history, they were reminded that their struggle for political freedom did

*not begin in the 1950s and '60s but had roots stretching
back to the days of slavery. They were also reminded that
their struggle for political justice in the United States had
always been associated with their churches. Whether in
the independent northern churches African American
Episcopal (AME), African Methodist Episcopal Zion (AMEZ),
Baptist, etc. or in the so-called invisible institution among
slaves in the South (which merged with the independent
black churches after the Civil War) or as members of white
denominations, black Christians have always known that
the God of Moses and Jesus did not create them to be slaves
or second-class citizens in North America.[2]*

One of the most prominent African–American denominations
is the African Methodist Episcopal Church. Founded in 1816 by
Richard Allen, the AME church now claims well over 2 million
members. The story of the founding of the AME church begins
with persecution and blatant racism on the part of Methodists in
Philadelphia.

*The AMEC (African Methodist Episcopal Church) grew
out of the Free African Society (FAS) which Richard Allen,
Absalom Jones, and others established in Philadelphia in
1787. When officials at St. George's MEC pulled blacks off
their knees while praying, FAS members discovered just
how far American Methodists would go to enforce racial
discrimination against African–Americans. Hence, these
members of St. George's made plans to transform their
mutual aid society into an African congregation. Although
most wanted to affiliate with the Protestant Episcopal
Church, Allen led a small group who resolved to remain
Methodists. In 1794 Bethel AME was dedicated with
Allen as pastor. To establish Bethel's independence from
interfering white Methodists, Allen, a former Delaware
slave, successfully sued in the Pennsylvania courts in 1807
and 1815 for the right of his congregation to exist as an
independent institution. Because black Methodists in other*

middle Atlantic communities encountered racism and desired religious autonomy, Allen called them to meet in Philadelphia to form a new Wesleyan denomination, the AME.[3]

The name of the organization bears some explanation. African refers to the fact the founders of the denomination were of African descent—not that it was founded in Africa, nor should it imply participation limited to peoples of African descent. Methodist refers to the obvious connection to the Methodist church, and the AME still views itself as part of the Wesleyan or Methodist movement. Theologically the AME is very close to the mainstream Methodist. Episcopal does not refer to the Anglican church or the Church of England, which is often called the Episcopal church in the United States. Rather, the term Episcopal refers to a form of church government, where the leaders of the denomination are Bishops of the Church.

At the time of this writing, there are nine denominations that are recognized as African American: the African Methodist Episcopal (AME) church; the African Methodist Episcopal Zion (AMEZ) church; the Christian Methodist Episcopal (CME) church; the National Baptist Convention, USA, Incorporated (NBC); the National Baptist Convention of America, Unincorporated (NBCA); the Progressive National Baptist Convention (PNBC); the Church of God in Christ (COGIC); The National Missionary Baptist Convention (NMBC); and the Full Gospel Baptist Church Fellowship (FGBCF). The FGBCF does not refer to itself as a denomination.

The majority of these denominations came out of the Holiness movement and have close ties with Pentecostalism.

End notes

1. Orthodox Christian Information Center, www.orthodoxinfo.com/inquirers/charmov.aspx, accessed 10/09.
2. Cone, J. H. (1984). *For My People: Black Theology and the Black Church.* Orbis Books.
3. www.ame-church.com/about-us/history.php, accessed 9/09.

ACOUSTICS OF THE EXPERIENTIAL STYLE CHURCH

We have examined the Celebratory style and recognize centrality of the Eucharist and the role that the architecture and acoustics play in the celebration of the Holy Mystery. In the evangelical style, the preaching of the Word must be supported by the acoustics and the architecture. We now examine the Experiential style of worship. As we have seen in Chapter 13, there are two forms of experiential worship, the Pentecostal and the Seeker Sensitive. There are some important differences with respect to optimal architecture for each, but the acoustical requirements are very similar. When the elements of experiential worship are examined, it is undeniable that music emerges as a very important element.

Music in Experiential Worship

A recent unscientific survey of students at a Christian college indicated that they equate music with worship. This perspective is supported by worship leaders who proclaim from the pulpit that the congregation is now entering a time of *worship* when actually what is being introduced is a time of *music*. This may be problematic from a theological perspective, but we will resist the

Sound of Worship. DOI: 10.1016/B978-0-240-81339-4.00015-6

temptation to weigh in on this one! Suffice it to say that especially in the Experiential style of worship, music is at the core of the experience. The root of this emphasis on music goes back to the Reformation where the reformers, in their zeal to empower the congregants and shift the power base away from Rome, taught the priesthood of all believers.

Before the Reformation people went to watch the priests perform the Mass. After the Reformation people were expected to participate and music was one of the best ways to promote participation. The Pentecostals recognized this, of course, and in the early days of the Pentecostal movement they actively encouraged individuals to start songs with the hope that they would catch on and be picked up by the congregation as a whole. They took seriously the instruction from the Apostle Paul, "When you come together each one of you has a hymn, a lesson, a revelation, a tongue, or an interpretation."[1] According to Calvin Johansson in his monograph, *Music in the Pentecostal Church*,[2] music in the Pentecostal church evolved first as a reaction to the establishment. Whatever the traditional church did, the Pentecostals did *not* do. If mainstream music was formal and scripted, the Pentecostals adopted spontaneity. "Their impassioned Pentecostal fervor, a newfound relationship with Jesus, a profound love for God along with the infilling of the Holy Spirit, left precious little tolerance for contemplating let alone using, 'the old way'."[3] Second, Pentecostal music was functional and pragmatic. If it achieved the desired results, it was used. Again according to Johansson, "If a Pentecostal musician were asked, 'What is the function of music in worship?' the likely answer would be, 'Music is a vehicle through which we praise God.'" But further answers would be sure to follow: "It can cover up noise, fill silence, create atmosphere, be a means of service, accompany singing and manipulate and control people. Church music, functional art that it is, has the ability to do many things."[4] It is not unusual for a Pentecostal pastor to encourage the musicians to play on, as it creates a mood. Johansson goes on to trace the evolution of Pentecostal music into the twentieth century, and notes that the manipulative character of the music was refined and honed into a high art.

Over time the spontaneity gave way to a much more scripted experience. "Standing people were admonished to praise and clap their hands along with rhythmic body swaying and whatever singing might take place prompted by an ever present video screen. The song service's function was to make people *feel*, which music is readily able to do. Objective theological biblical connections (as found in Hymns, for example) were abandoned for the emotional euphoria engendered by endless repetition of CCM (Contemporary Christian Music) songs. Part of the technique for accomplishing this had to do with the raw psychological manipulative power that extremely loud rhythmic music has over the nervous system. As the century progressed, trap set and microphone teamed up to make Pentecostal music an overt exercise in congregational control."[5]

This perspective is shared by the seeker-sensitive church as well, but perhaps not as bluntly stated. In his book, *The Purpose Driven Church*, Rick Warren writes about the power of music.

> *I'm often asked what I would do differently if I could start Saddleback over. My answer is this: From the first day of the new church I'd put more energy and money into a first-class music ministry that matched our target. In the first years of Saddleback, I made the mistake of underestimating the power of music so I minimized the use of music in our services. I regret that now.*[6]

Rick Muchow, the music minister at Saddleback, quotes Aristotle as saying, "Music has the power to shape a culture." He goes on to say, "There is no doubt that God is using music as a primary event feature in today's western culture. Connected to the Spirit of God, music is the most powerful tool available to reach and win your target. The Senior Pastor who believes this fact is secure in himself and will feel no threat when the congregation compliments or criticizes purpose driven music. The Music Minister who uses music strategically to serve and support the purposes of the church, as understood by the Senior Pastor, will enjoy the trust of the same."[7]

Acoustics

With music being as important as it is in this form of church, it is important that the building be designed around music. It is strongly recommended that the design of the experiential church be a cooperative venture between an architect and an acoustical engineer. These rooms need to be aural spaces rather then visual ones. Especially in the seeker-sensitive church, it is highly unlikely that there will be any form of visual symbolism. This is not to suggest that these rooms be designed as concert halls. The experiential church requires a performance space, which has more in common with a movie theatre than a concert hall. The music will all be amplified and high impact. The system needs to be capable of delivering levels in excess of 110 dB SPL. That is not to say that the levels in any church setting should be that loud, but the system needs to be able to deliver that kind of level if there is to be enough headroom. Acoustically, this demands that the room be as dead as possible; that is, the mid-band RT_{60} should be very low. A slight bass multiplier is tolerable as long as it is under one second. A bigger problem is building an acoustical shell that will contain and not absorb the bass. If gypsum board will be used, especially if there are large surfaces area to be covered, there needs to be at least two layers on 16" o.c. studs. If only one layer is used or wider stud spacing is used, there will be excessive low end absorption and it will be very difficult for the sound system to produce the impact that is desired. CMU (Common Masonary Units; i.e., cement block) or brick are the preferred materials for the sanctuary.

Focusing surfaces must be avoided. The rear walls need to be as absorbent as possible so that the sound from the sound system does not reflect either back to the stage or to the congregation. There needs to be absorption on stage as well in order to help control the sound from the stage monitors. The one surface where there may be a difference between the seeker-sensitive experiential church and the Pentecostal derivative church is the ceiling. In the seeker-sensitive church, congregational or participatory singing is not a big deal. It is assumed that the "seeker" will probably not know the songs and would not be inclined to sing anyway about an experience he or she may not have had. The ceiling can be very effectively used for absorptive panels or treatment. The Pentecostal church on the other hand, values audience participation. Even though the Pentecostal church should have a very low reverb time, and the actual deck should be absorptive, ceiling reflectors 15 feet or so above the congregation will allow the worshippers to feel like their participation matters.

For many experiential style churches, a choir is an important part of the worship experience. Choirs are a challenge for a number of reasons. First, it is a challenge to mic a choir. If the microphones are placed close to individual singers, the sound system does not have to work as hard, but you lose the sense of the whole. If you back the microphones away from the singers you can capture more of the whole, but the sound pressure drops off and the system has to work harder to get the sound of the choir out to the audience. Some have tried to use large numbers of wireless microphones and attempt to mic a significant percentage of the choir. This requires not only deep pockets and very large consoles but sound systems with the necessary potential gain to allow large numbers of microphones to be active at one time. What many sound system operators do not seem to understand is that for any sound system/acoustic space combination there is a maximum amount of potential gain that the sound system can provide. That total gain can be dedicated to one microphone, or apportioned to 50 microphones. The total potential gain does not change (see Chapter 20).

The second challenge that choirs present is finding a way for them to hear themselves. This is sometimes attempted with elaborate monitor systems, but this also can cut into the total gain before feedback.

A better solution is to include a choir shell into the design of the worship space. A choir shell can be a very effective way to let the choir be heard and for them to hear themselves as well.

The Experiential church can be another good candidate for reverberation enhancement systems. The room can be built acoustically dead, and the enhancement system used to created the sense of a large and live acoustic space when needed.

The Repurposed Church

This is not a take-off on Warren's Purpose Driven Church! In urban areas especially, we are seeing some of the traditional churches closing their doors and selling their buildings. Sometimes the church building is bought by a new church, usually of a different variety. Sometimes these pairings can work. An Evangelical church can make do with the building of a former Celebratory church. A community style church can work with a building of a former Evangelical church. A Celebratory church might even be able to convert a former Evangelical church. The pairing that is the most difficult is when an Experiential church buys a former Celebratory church, especially a gothic, reverberant structure. Unfortunately this is not uncommon. A Pentecostal church that buys an old Catholic church has to understand that the cost of renovation is going to include a significant amount for acoustical treatment, and even then it is unlikely that the building will ever feel right.

Case Study

12Stone Church

12Stone Church's worship facility in Lawrenceville, Georgia, a suburb of Atlanta, is a visually stunning, entirely HD theatrical space laden with some of the newest broadcast-quality components. The space itself is not called a sanctuary; rather it's quite intentionally deemed a "worship experience center."

According to Michael Wright, president of TI Broadcast Solutions Group of Norcross, Georgia (who also served as principal designer for all the HD video systems at 12Stone), "The idea is to envelop people in the

Entrance to 12Stone Church.

room with world-class sound, lighting and, of course, HD video. Those three elements working together in concert produce an environment where ultimately the people in the crowd are invited into an intimate moment with God."

Weeks before moving into their 2,600-seat worship venue, the church stepped into a new season with the new name, 12Stone. Founded as Crossroads Church in 1987, the ministry presently serves more than 5,000 adults and children weekly.

The name 12Stone comes from the story of Joshua, who led the Israelites into the Promised Land. Before reaching the Promised Land, however, the Israelites needed to cross the flooded Jordan River. God told Joshua to have the priests step into the water. When they obeyed, the water was cut off, allowing the Israelites a safe crossing. On the rivers edge, 12 stones were stacked to represent the 12 tribes of Israel and the rescuing hand of God.

Interior of 12Stone Church.

Similarly, the vision of 12Stone is to "inspire bold crossings in the lives of everyone we come in contact with," states Kevin Myers, founder and senior pastor of 12Stone.

Design and installation of 12Stone's all-important audio system was provided by dB Audio & Video out of Gainesville, Georgia. Ronnie Stanford, dB system's analyst and front-of-house (FOH) mixing volunteer at 12Stone highlights the worship center's layout. "The room features a thrust stage. When the communicator stands at the end of the stage, there is a 230-degree radius around him. From that point there is no seat more than 100 to 110 feet away," notes Stanford. By adding only 10 feet of distance between the stage and the farthest seat in the center, the seating capacity more than doubled for the new room.

The loudspeaker system features products from Danley Sound Labs. Mike Hedden, 12Stone's audio system designer and president of dB Audio & Video, says, "The SH 50s were chosen for this room because of their arrayability and pattern control. This was important not only because of interaction between each individual loudspeaker in the cluster, but also in regard to the interaction between the clusters of the LCR design." Given the shape of the room and the thrust stage, selecting a loudspeaker with good pattern control was also important to help reduce the possibility of feedback caused by open microphones used in close

Danley SH50 Loudspeakers in Ceiling.

proximity of the loudspeaker system. 12Stone's main sound reinforcement system consists of 22 Danley SH 50 full range loudspeakers. "The SH 50 has excellent pattern control and seamless coverage across multiple arrayed cabinets," maintains Stanford. Three clusters of six cabinets (three over three) cover more than 70 percent of the room in a LCR configuration. Side fill clusters of two cabinets cover the extreme sides of the 220 degrees of seating around the thrust stage.

Four Danley Sound Labs TH 115 Tapped Horn Subwoofers are used to help extend the low frequency response of the full-range SH 50s. Nine SH 100 speakers, used as delays, are positioned at the rear of the room, complementing the fill speakers.

The SH 100 delays are integral for seating behind the suspended screens. Without them, these sections (six to eight rows deep) would be completely blocked by the screens. Thus, even coverage is ensured as a result of the delays filling in gaps at higher frequencies all the way to the back of the room—an important factor when dealing with speech intelligibility. "I think that even without the delays as far as music is concerned, you would be fine, but when it comes to the pastor speaking, the intelligibility of speech, you need those delays," remarks Stanford.

All speakers in the room are self-powered speakers. "... When you talk about 22 speakers, plus nine delays and five subwoofers, you would need a couple of substantial racks to handle all the amplifiers—it would be pretty intense. So, part of the design was to use self-powered speakers for their simplicity and ease of installation," says Stanford.

Primary control of the FOH system is with Yamaha's M7CL 56-channel digital console (48 mic inputs plus four stereo inputs). In fact, there are a total of four Yamaha M7 consoles used in other rooms on the campus. "Since our technical ministry is driven by volunteers, we decided it would be a wise decision to use the same console for continuity's sake. When you have volunteers running your technical equipment and they are doing it with incredible excellence— you have that continuity and ability to expand ministry without a huge learning curve," explains Stanford.

12Stone monitor world is a combination of the Yamaha M7 and an Aviom personal monitoring system. All musicians are on Sennheiser 300 series wireless in-ears with their mix created through the M7. Because creating a mix for wireless in-ears is a unique technique—a skill that is learned—the church brought in a professional audio mixer to help volunteers develop a template to create a proper mix.

Vocalists have the autonomy to create their own monitor mix through the Aviom. Stanford notes that the rack mount version of the mixers was selected to free the stage of additional gear. During rehearsal vocalists go backstage to set their mix once for services.

The other part of the audio system in the room is an electronic architecture acoustic enhancement system called LARES. "It's an amazing system," notes Stanford. "There are 58 Danley Sound Labs SH100 loudspeakers positioned around the room. They are flown such that they fire straight down, not at an angle." The source for the SH100s are four Schoeps microphones hanging over the perimeter of the audience. dB Audio & Video's Hedden adds, "The LARES system is important because it allows an acoustically dead space to become a controlled reverberant environment, which is unique to most contemporary churches. LARES gives us 18 dB of gain out of the box, and the SH 100s from Danley Sound Labs provide incredibly dense coverage, a combination that was critical in a room with fan noise generated by the five video projectors and other lighting components." Hedden worked with Steve Barbar of LARES Associates in Belmont, Massachusetts, on the design of the electronic architecture system. Stanford continues, "You create an acoustic environment that can be changed. If you clap your hands in this room [with the system off], it's just dead. But when I turn the system on and go to different settings, you move from a very light reverberation to feeling like you're sitting in the middle of a football stadium indoors—so its impact is huge. We use it during our worship time for congregational singing to make the room more live." Hedden feels that using LARES to make an environment more "live" provides an atmosphere that is more conducive to congregational singing and participation during worship.[8]

Suggested Listening
Listen to the LARES demo in the Reverberation section at www .sound-of-worship .com.

End Notes

1. Corinthians 14:26.
2. Johansson, C. (2007). Chapter 4, "Music in the Pentecostal Church". In: E. Patterson, E.J. Rybarczyk (Eds.), *The Future of Pentecostalism in the United States* (pp. 50–63). Lexington Books.
3. Ibid., p. 51.
4. Ibid., p. 56.
5. Ibid., p. 57.
6. Warren, R. (1995). *The Purpose Driven Church* (p. 279). Grand Rapids Michigan: Zondervan.
7. http://www.encouragingmusic.com/articles/MusicAtSaddleback.asp, accessed 4/10.
8. Excerpted from http://www.churchproduction.com/go.php/print_article/6083, accessed 7/2010.

COMMUNITY WORSHIP

16

THE COMMUNITY WORSHIP STYLE
Let Me Be as Christ to You

Introduction

The Community Worship style is by far the least common of the four worship styles. It shares many of the traits of others, especially the Evangelical style, but differs in emphasis in very important ways, especially in terms of the acoustics of the spaces and the use of media in the service. We are using the word *community* to mean that the focus and centrality of the worship experience is about doing it as a group, or *community*, of believers. The presence of the word *community* in the name of the church does not necessarily indicate a community worship style.

In these types of churches, where the building most often is referred to as a meeting house rather than a church, the congregants will feel like they have "done church" if they have felt like

Sound of Worship. DOI: 10.1016/B978-0-240-81339-4.00016-8

they were a part of the community of believers and were able to minister and be ministered to. In these congregations, the idea of the members "being Jesus to each other" is well developed and in no way seems sacrilegious. The Church of the Brethren, one of the groups who typically worship the community style, have as their motto, "Continuing the Work of Jesus. Peacefully. Simply. Together."

Those who are most likely to adopt this form of worship are those who are descendant from Anabaptists, including the Mennonites, the Brethren, the Amish, and others. There are some churches who practice a community style of worship but are not historically related to the Anabaptists, most notably the Moravians (see Chapter 9). We will be examining the Anabaptist theology and ecclesiology as it is their way of thinking about God and church that leads them to this worship style. Their theology and praxis is much more a horizontal one, in which their devotion to God and to some degree their understanding of salvation is demonstrated and worked out through their relationships with their fellow man. In his book on Anabaptist ecclesiology, *Building on the Rock*,[1] Walfred Fahrer starts out with an analysis of the passage in Matthew 16 where Jesus and Peter are talking. Jesus asks Peter, "Who do men say that I am?" Peter answers that some say one thing, others something else. Jesus then puts the disciples

Figure 16.1 First Mennonite Church, Bluffton, OH.

on the spot and asks them straight out, "Who do you say that I am?" Peter responds, "You are the Christ, the Son of the living God." And Jesus said to him, "Blessed are you, Simon Barjona, because flesh and blood did not reveal this to you, but My Father who is in heaven. I also say to you that you are Peter, and upon this rock I will build My church; and the gates of Hades will not overpower it."[2]

Fahrer shows three different ways that this passage has been historically interpreted. The first way is to suggest that the rock that Jesus refers to is Peter himself. Jesus is making a play on words as Peter or *Petros* is Greek for stone or rock. Fahrer quotes a Roman Catholic Decree, *Laetentur Coeli* 1439, "We define that the Holy Apostolic See and the Roman Pontiff (the Pope) have primacy over the whole world, and that the same Roman Pontiff is the successor of Saint Peter, prince of the Apostles, the True Vicar of Christ, the head of the Church."[3] What the Catholic church is saying is that Jesus built the Church on Peter, meaning that Peter was the first leader of the Church, essentially the first Pope. Since according to tradition, St. Peter was the founder of the Church in Rome, the Roman church maintains that it has a sort of primacy over the other churches. This question of primacy was at the heart of the great schism of 1054 and was covered in Chapter 5. Authority over the Church then is passed down through the generations to each of the Popes, who view themselves as successors to St. Peter. This belief or interpretation resulted in an organization that remains unequaled in history and Church as sacred ceremony.

The Matthew 16 passage can be understood in other ways, however. Fahrer quotes Martin Luther, who challenged the Roman position on many issues including authority and leadership. Luther writes, "The Lord then says, 'And I tell you, you are Peter and on this rock I will build my church.'… Now the Lord wants to say, 'You are Peter, that is, a man of rock. For you have recognized and named the right man, who is the true rock, as Scripture names him, Christ, On this rock, that is, on me, Christ, I will build all of my Christendom.'"[4] This is essentially the Protestant position, that the rock is Christ and the Church is built on Him. This correct assertion by Peter was in a sense the prototypical example of correct doctrine, which is the hallmark of the Evangelical church. Correct theology, learned from correct teaching, will result in correct living. The church that evolved from this position is a church that values and proclaims the Gospel.

Fahrer asserts that there is a third way to interpret this passage. He quotes Pilgrim Marpeck, an early Anabaptist from Germany, "On this *testimony* [emphasis mine] of Peter the new church of Jesus Christ was built... in accordance with the words of Christ when he says: 'Upon this rock I will build my church or congregation' (Matthew 16:18)... Further he says: 'to this church I have committed the keys of heaven which forgives sins and retains them' (Matthew 16:19-20)."[5]

The Anabaptist position is that it is the *confession* that is the rock. They maintain that Jesus did not simply congratulate Peter for getting it right, nor did he reward Peter for this insight by making him Pope. Jesus said to Peter that God the Father revealed this to him, as a result of a step of faith. So the confession *as an act of faith* is the foundation for the church: "...it is neither Peter's legacy nor the doctrinally correct statement which is the foundation of the church, but the genuine confession of faith in Jesus..."[6] The result of this conviction is church as a community of believers, not built on historical tradition or on doctrinal orthodoxy, but on relationships based in faith. Fahrer states, "We are closer to the vision of Jesus when we understand church as a community of believers who have voluntarily declared their faith in Christ as Lord. The Church is a faith community."[7]

This emphasis on community is at the root of many of the beliefs and practices of the Anabaptist way of "doing church." One core belief for many Anabaptists is a very high regard for the sanctity of life. This belief finds expression in a number of ways.

Anabaptists are most often pacifists who value all human life equally and will oppose war, abortion, and capitol punishment with equal vigor. A popular Anabaptist sentiment sometimes seen on bumper stickers is, "When Jesus said love your enemies, He probably did not mean kill them."

ANABAPTIST PACIFISM

The Anabaptist position on pacifism dates back to the very beginnings of the Anabaptist movement. In 1632 there was a Mennonite Conference held at Dordrecht, Holland, which produced what has become known as the Dordrecht, Confession. It stands as one of the early expressions of Anabaptist theology and ecclesiology. Article XIV titled *Of Revenge* states in part:

> *As regards revenge, that is, to oppose an enemy with the sword, we believe and confess that the Lord Christ has forbidden and set aside to His disciples and followers all revenge and retaliation, and commanded them to render to no one evil for evil, or cursing for cursing, but to put the sword into the sheath, or, as the prophets have predicted, to beat the swords into ploughshares (Matthew 5:39, 44; Romans 12:14; I Peter 3:9; Isaiah 2:4; Micah 4:3; Zachariah 9:8, 9).*
>
> *From this we understand that therefore, and according to His example, we must not inflict pain, harm, or sorrow upon any one, but seek the highest welfare and salvation of all men, and even, if necessity require it, flee for the Lord's sake from one city or country into another, and suffer the spoiling of our goods; that we must not harm anyone, and, when we are smitten, rather turn the other cheek also, than take revenge or retaliate. Matthew 5:39. And, moreover, that we must pray for our enemies, feed and refresh them whenever they are hungry or thirsty, and thus convince them by well-doing, and overcome all ignorance. Romans 12:19, 20.*[8]

Down through the centuries pacifism has been one of the benchmarks of Anabaptist teaching and it remains an important part of what it means to be Anabaptist.

Typical Community Worship Service

Because of the emphasis on relationships and community, it is difficult to describe a typical Community worship service.

On the surface it may look very similar to the evangelical style. There may be the usual elements of prayer, music, the Lord's supper or communion, and teaching, but the emphasis will be markedly different from the other styles.

Prayer

In this style of worship, prayer has the same meaning as it does for the Evangelical; it is talking with God. However, in these churches the prayers are more likely to be supplications on behalf of members of the church community in crisis, or on behalf of the poor or disenfranchised in the wider community. It is not unusual to hear prayers of supplication asking God to bless our enemies, meaning individuals with whom we may have unresolved issues as well as the enemies of the state. When there are acts of violence, Anabaptists will pray as much for the perpetrators of the violence as for the victims.

Music

In the Community church tradition, music has a checkered past. Historically there were those of this tradition who prohibited musical instruments of any sort in the meeting house. Those who held this view often had very well-developed congregational singing with four or more part harmony. Others allowed instruments in the meeting, but the role was always as accompaniment. The notion of performance music would be out of place because the emphasis in the Community church is never on the individual but rather on the collective. Four-part a cappella singing is, for many in this form of worship, an exemplar of the community of believers forming the Church. In the book, *Sound in the Land: Essays on Mennonites and Music*, the author of the last chapter (Laura H. Weaver, a 73-year-old Mennonite) writes of the four-part a cappella singing as being a sort of survival strategy,[9] allowing her to experience community even when living outside of a traditional Anabaptist context. For her, community "is re-created each time I hear and participate in such singing."[10]

Many contemporary Anabaptists have adopted more complex forms of music into the worship service. Even though there may be instruments and contemporary forms, the emphasis will still be on community rather than performance. Anabaptist communities find it difficult to use contemporary music from other traditions as it is often very individualistic and emotional. These values are in tension with Anabaptist core values, which place the community above the individual.

Scripture Reading

In this style of worship, there will likely be reading of scripture. The Bible is an important part of the life of the Anabaptist community. They hold, as do the Evangelicals, that the "Bible is the book of the people of God."[11] All churches struggle with how exactly to understand and apply the Bible. Not surprisingly, the Anabaptist believes that it is not a hierarchy of authority, nor an individual, but the *community* that is responsible for the interpretation of the scriptures. "The Bible is to be interpreted within the context of the believing, obedient community, as that community seeks to communicate its message to the world. The authority of the Bible becomes evident within the covenant relationships of the believing community as it is led by the Spirit, who was sent forth by the victorious Christ."[12] It is the corporate interpretation, the commitment to working together to understand the scripture that protects congregations from the whims of charismatic leaders with unorthodox views.

Preaching

In this style of worship there will be a wide variety of kinds of preaching. Many churches that practice this style will refer to this activity as teaching rather than preaching. It will generally consist of exegesis of some scripture passage and usually some very practical application to the life of the community. Some will adopt a more evangelical approach and preach a sermon that may fit very well in an Evangelical church. The difference is again in emphasis. In this style of worship, the preaching/teaching is not the highpoint of the service, although it is an important part. Unlike in the evangelical style of worship there will rarely, if ever, be any sort of alter call or public invitation.

Other Elements of the Service

There is a lot of variance in many of the ancillary elements of community worship. The collection of tithes and offerings is a good example of the diversity found in this style. Those churches who maintain connection with their more traditional Anabaptist roots will not collect an offering per se. Collection boxes will be strategically placed near the exits of the meeting-house and may not even be mentioned during the course of the service. Others at the other end of the spectrum may actually pass offering plates.

Communion or the Lord's Supper is practiced by virtually all those who worship in this style, but again with a wide variety. Traditionally, Anabaptists rejected the idea of any sort of mystical activity associated with communion. They tended to side with Zwingli and saw it as purely commemorative. This traditional stance is still found in modern Anabaptist churches. "The Mennonite interpretation of Christian belief concerning the Lord's Supper has the following traits: It is opposed to speculation about metaphysical changes in the elements, yet often rationalistic in its arguments against such speculations. It has a spiritualistic tendency which is manifest in its inability to make positive claims for the relationship in communion between bread and the presence of Christ. At the same time the traditional piety of participants in communion indicates belief in a mysterious reality beyond what Mennonite teaching can explain. The Mennonite interpretation of communion has emphasized the human more than the divine action of grace. This has sometimes been a way of countering claims for divine action which Mennonites deem to be unbiblical. More often it has issued from the conviction that ethical response is the most profound act of gratitude for grace."[13] Communion may be observed weekly, monthly, or annually.

In these churches there will be a strong emphasis on the communal aspect of the Lord's Supper.

Often the elements will be served to the group by passing them from one to another rather than each receiving the elements from a priest or clergy. In this way the priesthood of each believer is emphasized. Because there is a biblical prohibition of participating in communion if there are unresolved issues between believers, Anabaptists would traditionally go to great lengths to make sure all the members of the community were in right relationships before communion was administered, even to the point of delaying the communion service until relationships were made right. This practice is still observed in some of the Amish communities. Not all community-based churches will take it to that extreme, but the community is highly valued and communion is seen as a way of preserving relationships.

Baptism is an important part of the Community church, although maybe not as important as it is for some of the Evangelical churches. It will see Baptism as symbolic rather then sacramental, an external sign of an inward act of faith. Again emphasizing the priestly role of all the believers, new converts often are baptized by members of the community who have mentored them in their faith rather than by clergy. The Community church will emphasize the necessity for a mature decision, rejecting the practice of infant baptism. Indeed this was the issue back in the late fifteenth century that gave birth to the Anabaptist movement. To more fully understand the Anabaptist we return to the fifteenth century and meet Menno Simons.

End Notes

1. Fahrer, W. J. (1995). *Building on the Rock: A Biblical Vision of Being church Together from an Anabaptist-Mennonite Perspective.* Herald Press.
2. Matthew 16:13-18 New American Standard Version.
3. Ibid., pp. 19–20.
4. Ibid., p. 20.
5. Ibid., p. 20.
6. Ibid., p. 22.
7. Ibid., p. 23.
8. http://www.bibleviews.com/Dordrecht.html#XIV, accessed on 10/09.

9. Epp, M., & Weaver, C. A. (Eds.), (2005). Sound in the Land: Essays on Mennonites and Music (p. 210). Kitchener, Ontario: Pandora Press.

10. Ibid., p. 215.

11. Mennonite Church, Biblical interpretation in the life of the church (Mennonite Church, 1977), *Global Anabaptist Mennonite Encyclopedia Online*, 1977, retrieved 20 November 2009, http://www.gameo.org/encyclopedia/contents/B5383.html

12. Ibid.

13. Krahn, C., & Rempel, J. D. (1989). Communion, Global Anabaptist Mennonite Encyclopedia Online. retrieved 20 November 2009, http://www.gameo.org/encyclopedia/contents/C654ME.html

17

THE ANABAPTISTS

Menno Simons (1496–1561)

To understand the Anabaptists, we will examine a fourth part of the Reformation, the so-called *Radical* Reformation. Luther, Zwingli, and Calvin were part of what has been called the Magisterial Reformation as their movements did not totally alienate the state. Indeed, many of the reformed churches were replacing the Catholic church as the approved state church and the followers of these non-Catholic churches often had some degree of protection from the state. The *radical* reformers, however, found themselves at odds with both the church and state. One of the early Anabaptist leaders was Menno Simons (Figure 17.1). Born in an area now part of the Netherlands, Simons was a son of poor dairy farmers. He studied to be a priest in the Catholic church, and was ordained into ministry in 1524.[1] In the 1520s a debate arose in Protestant circles regarding the question of baptism. Since Zwingli and others questioned the validity of the sacraments and tended to think in terms of symbolism rather then actual merit, the rite of baptism was fair game for debates. A growing number of Protestants were in favor of adult baptism as a conscious act of the will, engaged in by a believer who wished to publicly identify with the church.

This would have meant *re-baptizing* the adult since virtually every child was baptized as an infant into the Catholic church.

Sound of Worship. DOI: 10.1016/B978-0-240-81339-4.00017-X

Figure 17.1 Menno Simons, (1496–1561).

Those who practiced this re-baptism became known as the Anabaptists, *ana* being the Greek prefix meaning "re" or "again." This may seem innocent enough, however during the fifteenth century, two heresies were punishable by death. One was denying the existence of the Trinity, the other was being re-baptized! The Roman Catholic church taught that baptism was a sacrament that was effective for securing a place in heaven in the afterlife. To be re-baptized implied that the first baptism performed by the Church was ineffective. This was a position that the Church could not tolerate. The mainline reformers like Luther and Calvin had no problem with the Catholic practice of infant baptism although they had a different understanding about the efficacy of the sacraments. From the Roman Catholic perspective, Luther et al. were heretics and should be excommunicated from the Church. Their crimes and heresies, however, paled in comparison to those of the re-baptizers!

There were two events in Menno Simons' life that were instrumental in his leaving the Catholic church and becoming an Anabaptist himself. First, during his early days as a priest, he began to question the Catholic church's position that during the Mass the bread and wine actually miraculously changed substance and became the actual body and blood of Christ. To try to make peace with his inner struggle, Simons searched the New Testament for support of the Church's position. Finding none, he was faced with a very difficult decision. What constituted the final authority, the Church tradition or the Scripture? This indeed was the same struggle the other reformers had faced as well. Simons continued to serve in the Church, but he realized that in so doing, he was betraying his own convictions. This duplicity weighed heavily on Simons but he did not yet break from the Church.[2]

The event that ultimately would cause Simons to completely break with the Church occurred in 1531. A Dutchman by the name of Sicke Freerks was beheaded by the Church for being re-baptized during the previous year. Simons was appalled as this was his first experience of the Church putting someone to death for this heresy. Simons sought out the writings of Luther and other reformers but found no answer there as they basically upheld the Church's position on baptism. So again Simons went to the Scripture for an answer. He searched the Scriptures but found nothing to support the position of the Church. Still he stayed on in the Church, administering the Mass and baptizing infants, all the while realizing that in his heart he did not agree with the very acts he was performing. This spiritual struggle lasted about five years, and finally in January of 1536, Menno Simons ended his double life and publicly renounced the Church he had faithfully served his whole life. Unlike Luther, Zwingli, and Calvin, who all held positions regarding baptism that did not place them on the wrong side of the line, Simons was now not only a heretic but guilty of a capitol offense against the Church, and as such could be executed for his beliefs. He spent much of his time in hiding, but was recognized as a leader among the Anabaptists of Holland.[3]

The Anabaptists stood firm and many hundreds of them were tortured and executed for their beliefs by the church of Rome and by the emerging Protestant state churches as well. The Anabaptists were persecuted for their beliefs and practices regarding baptism, but many of them also believed that since Jesus taught nonviolence, His followers should be nonviolent, and they refused to bear arms for the state. The Anabaptists were really the first group who held that church and state should be separate. They believed that the Church was a higher power than the state, and that followers of Christ owed allegiance to no man, but to God only. This created a double jeopardy that made life very difficult for the Anabaptists. Not only did Rome view them as heretics of the highest order, but the state viewed them as traitors. No one really knows how many were tortured and killed for their beliefs. In 1660, a Dutch Mennonite Pastor published a book called *Martyrs Mirror of the Defenseless Christians*. This work, which is now in the public domain and online, gives the accounts drawn from public records of thousands who were put to death for their faith in the sixteenth century.[4]

Unlike some of the other reformers of the day, the Anabaptists did not develop complex theological statements. They were simple folk, farmers mostly, who believed that their faith in God would be manifest in their relationships to their fellow man. This aspect of Anabaptist beliefs can still be seen today in the emphasis on community. Menno Simons died in 1561 of natural causes; however, many of his close friends and supporters are listed in *Martyrs Mirror*, having paid the ultimate price for their faith. Followers of Simons' teaching became known as Mennonites, a group that exists today. Others who trace their roots to the Anabaptist are the Amish, the Hutterites, the Brethren, and some of the Baptists.

End Notes

1. Historical notes from the Church of the Brethren Network, http://www.cob-net.org/text/history_menno, accessed 8/09.
2. Ibid.
3. Ibid.
4. http://www.homecomers.org/mirror/intro.htm, accessed 8/09.

18

THE ACOUSTICS OF THE COMMUNITY CHURCH

CHAPTER OUTLINE

In Celebratory church we have sacred space where every element is there to provide a context for the celebration of the Eucharist. In the Evangelical church, the Word is preeminent and the acoustics must be optimized for intelligibility. In the Experiential church it is the total experience that informs the acoustic and sound system choices. Finally, in the Community style of worship it is most important that the congregation needs to be able to hear themselves and feel as though they are a part of the whole.

As mentioned in Chapter 16, an examination of the elements present in the Community style reveals all the same elements that are present in the Evangelical church. The important difference is one of emphasis. In the Evangelical style everything is there to support that preaching of the gospel. In this style, the building must serve the community. Traditionally in this style, especially among those who are from Mennonite background, the building is referred to as a meeting house rather than a church. Historically this practice dates back to the sixteenth century when the Anabaptists were severely persecuted by virtually everyone and public meetings were very dangerous. Meetings were held

Sound of Worship. DOI: 10.1016/B978-0-240-81339-4.00018-1

in private homes and the home designated for a specific meeting was referred to as the meeting house.

The largest of the Mennonite conferences or groups of individual congregations is the Lancaster Mennonite Conference. Started in Lancaster County, Pennsylvania, the Lancaster Mennonite Conference now spans the entire east coast of the United States. In 2005 there were 17,496 members in 186 congregations this averages out to around 94 members per congregation.[1] This averages out to around 94 members per congregation. Church of the Brethren congregations are sometimes larger but it seems to be an attribute of the descendants of the sixteenth century Anabaptists that the congregations remain small. Perhaps it is due to the emphasis on community. Many Mennonite churches still meet in homes but when these churches reach critical mass so that they need a building there seems to be a tendency to refurbish an existing space rather then to build on a piece of land. There is also a tendency for churches to split and form new groups rather then to grow to the point where community life is difficult. Later on in the case studies at the end of this chapter we will examine three churches from this style, two that were purpose-built, and one rehab.

Architecture of the Community Church

We might assume that churches whose worship style emphasizes community would favor architecture that would emphasize community. We might expect to see churches in the round for example. Indeed there are some good examples of Mennonite churches in the round. We will be featuring two such Mennonite churches as case studies. Reba Place Church in Evanston, Illinois is an outgrowth of Reba Place Fellowship, a Christian community started in 1957. In the late 1970s they bought and renovated an auto body paint shop to provide a place for common worship. They chose to create seating in a semicircle with the possibility of extending the risers all the way around into a complete circle. The stated purpose for setting up seating in this fashion was to essentially force people to face their neighbors. "It is hard to carry unresolved resentments if you have to sit face to face with your neighbor in the meeting house."[2] The same was true of Living Water Community Church in Rogers Park, Illinois, and College Mennonite Church in Goshen, Indiana.

Surprisingly, however, there are very few examples of Anabaptist churches in the round. Relatively little has been written about the history of Anabaptist/Mennonite church architecture. In the early days of the Anabaptist movement, purpose-built structures were out of the question. Meetings were held in secret, most often in homes. This suited the early Anabaptists quite well since for them the church was the community, not the building. It is interesting that in the few documented cases where architecture was debated Mennonites often wound up with fairly gothic buildings. In April 1999, the *Mennonite Quarterly Review* published an article called "Mennonite Debates about Church Architecture in Europe and America: Questions of History and Theology." In this article Keith L. Springier writes of a few twentieth century debates over Mennonite church architecture. He concludes, "The history and debates about Mennonite and Puritan church architecture at seventeenth-century Amsterdam and twentieth-century America suggest, first of all, that Mennonite architecture must be observed in the wider picture of Protestant church architecture. It did not rise or continue in a vacuum. It borrowed from its surroundings, and at times others may have borrowed from Mennonites. Mennonites and Puritan nonconformists had a connection in church architecture at Amsterdam; and in later times Mennonites (for example, the Anabaptist reformers of the Bethel College Church) have looked at Puritan church architecture with approval. Some of the Puritan theories about architecture (as on simplicity and order) were relevant for Mennonites. Other Mennonites, however, refused to be bound by this austere tradition and borrowed from across the architectural spectrum. American Mennonites, especially west of the Mississippi, found the high-church Gothic style very attractive."[3]

Currently in the United States there are many different forms of Anabaptist churches. "There are two noticeable extremes among the Mennonites of America regarding church buildings. The ultra-conservative groups—the Amish, Hutterites, and others—have developed a principle that no special building is to be provided for worship purposes only. They meet in private homes or school buildings. What was originally a necessity has become a basic principle of their faith. The other extreme is that of those Mennonite congregations who, for what ever reason, accept almost any architectural patterns found in their respective communities. This has produced some very odd mixtures and contradictions like the towers found on many Mennonite churches of the prairies, which are remnants of fortresses of the Middle Ages."[4]

This is an odd juxtaposition of values. We may wonder if the community-based congregations, who meet in these buildings whose architecture stands in contrast to their values, are aware of the incongruity. Some are.

The Anabaptists are not the only group who practice the community form of worship. The Moravians (see Chapter 9) are a very small denomination in the United States. Their life together often centers around community. There is a Moravian church in a small town in the Midwest. Their core values are conspicuously displayed on a poster in an anteroom attached to the main worship space. The very first value on the poster is community. However, the architecture of the church building says nothing about community. Looking at the exterior, the steeple tower dominates the rectangular building, which houses the sanctuary. It is a simple structure fitting in with the rural surroundings, but it is reminiscent of a Baptist church or a church with a strong hierarchy like Presbyterian or perhaps even Episcopal. Inside, there are pews in rows, all parallel to the raised stage. There is a pulpit conspicuously in the center of the stage with a rail in front, as if to separate the congregation from the clergy. The pastor was asked if the architecture of the church building fit the core values of the church community. Her answer was immediate and forceful: "Absolutely not!" She recounted that the building was built some years ago by an architect who was a Presbyterian and who included those elements that he felt were important for worship. She longed for a structure that would reflect the values of the congregation. She would have liked seating in the round or at least in a semicircle, which would remove the "paternalistic division" (her words) between clergy and laity.

In terms of acoustics, the governing aesthetic should be intimacy. Fortunately, as we have seen, churches that worship in the community style are rarely large. It can be challenging to create a sense of intimacy in very large rooms. Generally, intimacy is the result of placing reflecting surfaces fairly near the congregations. Low ceilings and relatively small rooms will all result in a short initial time gap and contribute to the impression of an intimate space. Intelligibility is also a concern because there is always a teaching and a significant amount of various sharings in this style of worship. However, intelligibility in small, intimate, nonreverberant spaces is rather easy to accomplish. Unless the ceiling is too low, a single point system is ideal as long as coverage is adequate.

Case Study

White Oak Church of the Brethren, Penryn, PA

Acoustical Consultant: Dale Shirk

White Oak Church of the Brethren is a large conservative Brethren congregation. Their music consists entirely of a cappella congregational singing. Their building, built in the 1980s, is a wide fan shape 74 feet long and just over 100 feet wide at its widest. Plus there are overflow wings on the rear corners and a small balcony over the rear lobby. The main sanctuary seats about 900, plus several hundred more in the overflow and balcony areas.

The ceiling is 26 feet high down the center of the room, sloping down to the sides and rear. All walls are drywall, the ceiling is plaster, the floor has fairly heavy carpet on concrete, and the pews are padded. Reverberation time is fairly short, under 1 second mid-band, somewhat longer at lower frequencies.

The dominant acoustic feature in this room is the center section of the ceiling, which is a concave cone segment running front to back. The ceiling height is uniform front to back down the center of the room. At the front the radius of the curvature is about 7 feet increasing to maybe 50 feet or more, a wide shallow curve at the rear. This means the imaginary centerline of this cone starts high up on the front wall, and falls through the floor at about the fifth or sixth pew, and is way below the floor at the rear wall. Beyond the cone segment the ceiling is a sloped plane surface.

The architecture of the ceiling makes it impossible to locate a loudspeaker directly overhead. They have been through several. Although the height and angles could allow even coverage from a center cluster, the acoustic focusing of the ceiling couples the loudspeaker to the centered pulpit and lectern microphones, causing severe feedback problems.

Our sound system solution was to locate the loudspeakers to the sides, beyond the curved ceiling section. The main loudspeakers are Electro Voice HP9040 large format horns, set nearly flush into the front wall, with the throat section protruding back into the attic area beyond. A 1 × 12 inch woofer box was mounted near each horn at the front wall-ceiling junction. Additional Frazier CAT-35 coaxial 8 inch speakers are located as fills for the front corners and as fills for the balcony and overflow areas.

White Oak Church of the Brethren.

A QSC DSP-322ua 8 $\times$ 8 DSP handles all processing including functioning as an optional automixer for a few select microphones. Also the project included relocating the booth from the balcony to the main floor, and replacing the aging console.

Although generally we do not recommend doing so, in this case we found it beneficial to run the main speakers in opposite polarity to each other. Since they are symmetrical around the centerline where the pulpit microphone is and the centerline of the focused reflections, the resulting cancellation did give a few more dB of gain before feedback.

One other unusual item in this system was the prayer microphone. Since they routinely kneel for prayer, they had previously installed a microphone below the pulpit top to better pick up the preacher while he was kneeling. We kept the concept and the homemade security mount, but upgraded the microphone.

The client has been extremely pleased with the result, especially the increased gain before feedback, allowing even timid talkers to be clearly heard.

Report by Dale Shirk

Case Study

Salford Mennonite Church, Salford, PA
Electro-Acoustic Consultant: Shirk Audio and Acoustics

Salford Mennonite Church is the oldest Mennonite congregation in its area. Their present building was built in the late 1800s and renovated many times since. Their worship style includes a cappella congregational singing, and sometimes piano accompaniment. Occasionally they will feature special music ranging to gospel, folk, or bluegrass styles.

The sanctuary is a simple rectangle, 85 feet long and 53 feet wide. The ceiling is a curved barrel ceiling at 17 feet high down the center aisle, curving down to under 14 feet at the side walls. The walls are plaster on brick, with a section of stone veneer at the front. The ceiling is hard plaster on wire mesh, the floor has thin carpet on wood, and the pews are mostly not padded. This room has a fairly live acoustic signature, with reverberation over a second mid-band, rising to over two seconds at low frequencies.

The sound system that was in place when we first evaluated the church consisted of six ceiling speakers with 5 inch drivers in tuned back boxes. This system had very poor coverage, with bright hotspots directly below the loudspeakers and poor intelligibility elsewhere. Measurements and modeling showed that even

Salford Mennonite Church.

a higher density distributed ceiling speaker system could not achieve adequate intelligibility due to the high level of reverberant energy. A center cluster was unacceptable due to aesthetics, and it would have required multiple locations to be workable.

The solution that was settled on was a pair of custom line source speakers located in the front corners of the room. These 8.5 foot tall columns each contain two Bohlender-Graebener RD-50 line source ribbon drivers in a horn/waveguide that is tight against the side wall to prevent the side wall reflection, and directs its output away from the pulpit area.

This provides very good intelligibility to the rear of the room and even coverage. Like all line source loudspeakers, they do not project the low frequencies as far as the high frequencies. To reduce this uneven frequency response characteristic, the existing ceiling loudspeakers were used to add in an appropriate amount of bass as needed at different distances, with delay to match into the main arrival. To reduce the slap echo off the rear wall, fiberglass barrel diffusers were added. The sound control booth was also brought down to the main floor out of its former cubby hole.

B-G Line sources at Salford Mennonite.

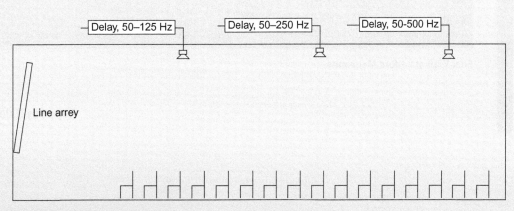

Delayed Low Frequency Enclosures.

The congregation was very careful in the purchase of this system, taking it a step at a time, requesting and paying for live demonstrations along the way. They had stringent requirements but were willing to pay what the proper solutions cost. The result is that they have intelligible speech reinforcement in this room for the first time.

Rear Wall of Salford Mennonite.

Report by Dale Shirk

Case Study

Reba Place Fellowship, Evanston, IL
Electro-Acoustic Consultant: EASI

Reba Place Church in Evanston, IL, is an outgrowth of Reba Place Fellowship. The Fellowship is an intentional Christian community started in the late 1950s by a group of folks from Mennonite and Brethren backgrounds. They wanted to live in community, sharing their resources, including their incomes. Members of the fellowship participate in a common purse, each member contributing their entire income and then living off of a modest stipend. Any money that is left over is used for a number of ministries, including supporting a number of disabled people who otherwise would have to be institutionalized. In the early years members of Reba Place Fellowship worshipped together in various homes in the neighborhood. As their numbers grew, they started meeting in a small store front. By the late 1970s they had outgrown the store front and they purchased an automobile paint shop and converted it into a meeting house. Seating was very carefully laid out in a half circle of seats on risers with a clear area in the middle for drama and liturgical dance. The service in the round is very important to the community as it is a symbol of their commitment to being in community. Since the mid 1980s Reba Place Church is a distinct entity from the fellowship but still values the roots of community upon which it was founded.

The roof of the meeting house is a bow truss design. The trusses are quite low, only about 12 foot clear. The "bow" of the bow truss has been filled in with drywall.

The relatively low ceiling creates a wonderful sense of intimacy and very good support for congregational singing, all very important to Reba Place Church. Unfortunately, this arrangement (Figure 18.6) makes it impossible to install a central point source sound system. The first sound system was a distributed system of 12 inch co-axial loudspeakers in large back boxes installed into the drop ceiling. This system had very poor coverage, with hot spots directly under the speakers, and dead spots where it was very difficult to understand the spoken word. This system was replaced in the late 1980s with a distributed system of Styrofoam speakers that looked like ceiling tiles. They were actually an improvement in terms of coverage and intelligibility. The problem was that when the system was installed Reba Place Church did not anticipate that the music they would use in worship would change dramatically. They changed from being light folk-oriented to having a more up-tempo gospel emphasis. The Styrofoam loudspeakers could not handle the dynamic range, so a new system was desired. EASI was retained to provide a performance specification, to review bids, and to perform the final voicing of the system. The new system consists of five full range boxes all delayed back to the center of the worship area. Provision was made for five separate monitor mixes as well as recording of the worship service. Intelligibility is measured at 0.86 STI and the coverage is +/−3 dB over the entire seating area from 100 Hz to 8 kHz.

Interior of RPC, Showing Ceiling, Speakers, and Seating Arrangement.

Report Courtesy EASI

End Notes

1. Global Mennonite Encyclopedia Online, www.gameo.org/encyclopedia/contents/L315, accessed 3/2010.
2. Private correspondence, Lehman, C. Seniro Castor, Reba Place Church, 2010.
3. Sprunger, K. (1999 April). *The Mennonite Quarterly Review.*
4. Krahn, C., van der Zijpp, N., & Kreider, R. S. (1989). Architecture, *Global Anabaptist Mennonite Encyclopedia Online*, retrieved 20 November 2009, http://www.gameo.org/encyclopedia/contents/A731.html

5

RESOURCES

19

INTRODUCTION TO RESOURCES

The next chapter contains resources that apply to all churches regardless of worship style. It begins with an overview of what should constitute a properly designed sound system no matter what kind of church is under consideration. Chapter 20, an overview, is written explicitly for architects or church committee members who may know absolutely nothing about sound systems or acoustics. It is not intended to be definitive or scientific. The rest of the section is much more rigorous, reviewing some of the more important acoustical attributes that pertain to churches.

When using this resource section it is important to keep in mind that there is an important difference between subjective assessment and objective measurement. Indeed, one of the more important tasks in the field of acoustics is to discover links between the subjective and objective domains.

The story is told of an acoustician back in the mid 1980s who was asked to fix a recording control room in the south side of Chicago. The consultant and the studio owner were from very different cultures and had different musical tastes. The owner was concerned about the low end of the control room. Nothing sounded right. The owner and consultant listened to records for a few hours and agreed that the low end was not right. The control room certainly needed work! The control room errors were documented using the best measurement platform of the day (Time Delay Spectrometry), which confirmed the consultant's notion of just how bad the room was. Carpenters were called in to tear apart the room and make the necessary repairs. After the room

Sound of Worship. DOI: 10.1016/B978-0-240-81339-4.00019-3

was reassembled and the monitors reconnected, the consultant and the owner sat down to listen to the improved control room. The consultant was ecstatic, the owner apoplectic! Clearly something was not communicated. The problem was in a very different aesthetic with respect to low end even though they both used the same vocabulary to describe it! The consultant wanted to hear the kick drum sound like 30 pounds of raw meat hitting the pavement after falling 20 feet. The owner wanted the low end to be all enveloping with no discernible attack but a huge warm bottom end hug! In its original form, the control room did neither. The consultant fixed it by making it sound the way he thought it should, exactly opposite from what the client asked for! The breakdown was clearly one of communication. It is important to note that the high tech measurements were of no help. They confirmed the consultant's point of view in that he could see that the room had a bit of low end "smear." The analyzer had no way of knowing that the client actually wanted *more* of that, not less! Whenever measurements are taken the result is always numbers! Numbers are just numbers. They have no aesthetic or predictive value unless some sort of *analysis* is performed. Using a tape measure to measure a couch tells the user that the couch is 34 inches wide. It does not tell the user that the couch will fit through the door. That requires analysis of the numbers. This story can be further analyzed using the following approach.

In Figure 19.1 there are three domains or collections of ideas or concepts that pertain to an event. Let's look first at the subjective domain. The subjective or perceptual domain is where all those things that we perceive about the event with our various senses are collected. Our perception of hot and cold, taste, tactile sensations, smell, and of course all the aural experiences like pitch, timber, and loudness all are part of this domain. In the story about the control room, this is the collection of impressions that each participant had.

The objective domain is where all the things that we measure are collected. We measure time, distance, mass, temperature, work, force, speed, acceleration, frequency, spectrum, and

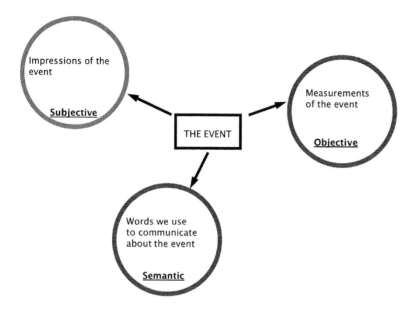

Figure 19.1 Three Domains.

sound pressure level to name a few. In our story this is the collection of TEF measurements that were used to document the problems in the control room. The objective domain is populated with the results of our observations of the event and those units of measurement that we have devised to try to quantify what we observe. It is interesting to note that we were not handed these units on some sort of divine platter. God did not directly speak to some prophet saying "behold thou shalt measure mass in this fashion." *We* made up the units and described the rules that we observed to the best of our ability. It is curious that in virtually no case does how we *measure* correlate with how we *experience*. When we double some quantity of something we do not experience the doubling as being twice as much. Two candles burning are not twice as bright as one candle burning. Doubling the power in an amplifier does not result in sound twice as loud, and so forth. We accept this as the nature of things. Yet in acoustics, as in other disciplines, this disconnect that exists between the measured and the perceived is a source of frustration. We wish that we could "map" the subjective into the objective and vice versa so that we could accurately and consistently predict one from the other. In our story the consultant wished that his measurements would give him the answers. This mapping is harder than it seems at first glance. Take a simple subjective impression like pitch. For most people, pitch is just another word for frequency. We might say that pitch is most often simply mapped over into frequency and indeed, frequency is the strongest predictor of pitch. It is not, however, the only predictor. It is easy to make people hear a pitch change without changing frequency.

There are at least four objective elements that play a role in the perception of pitch: frequency, amplitude, duration, and spectrum. So to fully describe the mapping of the subjective impression of pitch into the objective domain of measurement, you need at least these four components to accurately and consistently predict pitch from the measured (Figure 19.2).

As it turns out, pitch is one of the simpler ones. Subjective impressions like timber or sound space are considerably more complex in terms of their mapping back into the objective.

It is even more complicated than simply mapping between the objective and the subjective. There is a third domain—the

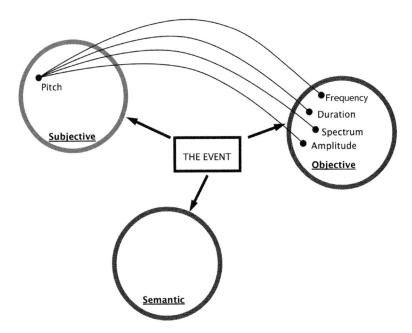

Figure 19.2 Mapping Pitch.

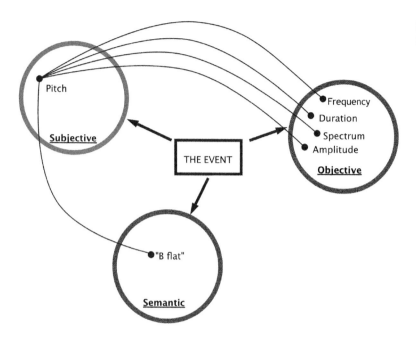

Figure 19.3 The Event Described.

semantic domain. This is the collection of words we use to describe both the objective and subjective domains (Figure 19.3). In our example of pitch, imagine someone with perfect pitch hears a synthesizer playing a sine tone at the pitch of B flat. She would describe the sound she heard as B flat.

The measurement of the event would yield numbers that would be reported using well defined terms, scales, and units (Figure 19.4).

The link between the objective and the semantic is obviously very straightforward. This is in essence the foundation of science. Science has very carefully defined most, if not all, of the objective quantities. It is pointless to debate the meaning of a kilogram of mass, or the length of a mile, or the definition of a volt. These things are pretty well cast in stone.

The relationship between the subjective and objective domains is very complex and not at all straightforward, for the obvious reason that everybody experiences events differently. However when we look at the relationship between the semantic domain and the subjective, we discover the most difficult mapping of all.

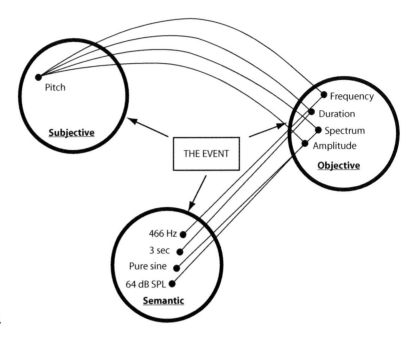

Figure 19.4 The Measurements.

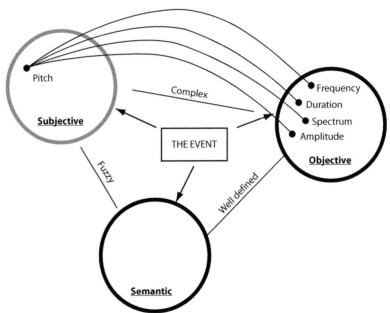

Figure 19.5 Summary.

The case we cited of someone with perfect pitch is a special case. Most people do not have the ability to call out exact pitches. But what if the event is something far more complex? The sound of rain? The sound of footsteps in gravel? The sound of a great cathedral? Not only does everyone have their own unique experiences, but since there are very few, if any, concrete or universally accepted ways of describing experiences, we often use different words to describe our experiences (Figure 19.5). Or as in the story told earlier, sometimes we use the same words to describe different subjective outcomes. It is the nature of language that some words and concepts are very concrete and the meanings unambiguous. To English speakers, water (noun) is understood by most everyone and it is not likely that someone asking for water will be given a cement block due to a misunderstanding. A record producer speaking to the recording engineer and complaining that the snare drum does not "grab" him is communicating a somewhat less clear message! The engineer has to figure out what "grab" means, then using devices that manipulate the signal in the objective domain, create something that is more pleasing or acceptable to the producer. Sometimes the words we use are part of a jargon understood by those "in the know," but not broadly understood. A guitar player asking an engineer for more "bite" on his sound will be understood if the engineer happens to be a guitar player, but probably not understood by the sound mixer if he is a piano player. Acousticians often get asked to fix "funny sounding" rooms. We cannot (or should not) simply respond in the subjective, but instead figure out what physical parameter of the room we can modify so that the experience is changed.

Acoustics has a remarkably scanty vocabulary. English is an extremely verbose language and yet there are very few words that describe an aural event. We are very dependent on onomatopoeia, metaphor, and borrowing from other senses to get the message across. "The guitar sounded 'bright' (vision) with a certain 'twang' (onomatopoeia) to it." Words that are strictly aural words like silent, quiet, din, cacophony, strident, sibilant, loud, most often describe either the *absence* of sound or some undesirable attribute. What vocabulary can we use to describe the majestic sound of a great cathedral? We discover all too quickly that there simply are no words. This makes the practice of acoustics rather difficult, especially with respect to trying to develop maps or links between the subjective and objective. We will see in Chapter 21 that there are terms like envelopment and clarity that have been defined at least in the context of concert halls. We have borrowed and in some cases redefined some of these terms for use in the church.

Finally, it is not the intent of this section to be a definitive work on acoustics. There are some very good acoustic texts in use in many academic institutions. We are covering key elements and you are encouraged to seek out more information when necessary.

To the architect, the message of the resources section is that worship is an aural event that is augmented by the visual. The architect is strongly encouraged to engage a competent acoustical engineer who will be sensitive to the needs of the church. This section will provide a primer in the language and concepts and enable the architect to share a common language with the acoustical engineer.

To the acoustical engineer/consultant there will not be very much if any new acoustic information. Rather, the message is that not all churches are the same and it is important to respect, understand, and support the ecclesiology and theology of each church when designing the acoustics and/or sound systems.

To the church building committee, this is a section of resources that should help you ask the right questions of the right people.

20

ELEMENTS OF CHURCH SOUND SYSTEMS

CHAPTER OUTLINE

Introduction: The Case For Systems Design

Note: This chapter is written for those who may know little or nothing about sound systems but need to be involved in decision making regarding sound systems. Architects and church facility

Sound of Worship. DOI: 10.1016/B978-0-240-81339-4.00020-X

committees will find this a useful resource. It is not intended to be a complete or comprehensive guide to sound systems design. Rather, it will help those who are serving the church ask the right questions, to help insure that the church gets the best sound system for their needs and for their money.

Every church large enough to require a sound reinforcement system, and whose ecclesiology allows for it, will have to purchase and install a sound reinforcement system at some point in its history. It has been noted that churches often go through two sound systems before finding something that works. The first system may have been installed when the building was built by the lowest bidder; or for churches built before sound systems were available, the first system was installed before sound systems design was any sort of discipline. This first system often never works correctly. The next system many churches go through is one designed by well-meaning members of the congregation who may have some experience in sound or electronics, but often lack the training or knowledge to design a proper system. System 2 is often only a slight improvement over system 1. Finally in desperation the church spends the money to "do it right," and hires a reputable systems designer/sound system installer who gets it right.

The church might legitimately ask, what does it mean to get it right? Does it really matter? For some churches is might not. For most it does. Don and Carolyn Davis published a sort of memoir of their many years with Syn-Aud-Con and professional audio. The book is called, *If Bad Sound Were Fatal, Audio Would Be the Leading Cause of Death*.[1] The title says it all! Possibly because there is "so little" at stake (i.e., nobody dies) or maybe because we are a predominately visual society, we often settle for inferior sound systems. In an audio–visual presentation, people will often accept as inevitable sound that is distorted, garbled, and all but unintelligible, but get indignant if the projector is out of focus. In the twenty-first century, there is just no excuse any more for bad audio. Truth be told, really high quality audio has been easily available for decades and it doesn't have to cost outrageous sums. The key is to get the right people involved.

Sound systems design is certainly a science and to some degree an art. Beware of so-called professionals who refuse to utilize any form or measurement or scientific inquiry. This may be sound system specification, but hardly rises to the level of design. A properly designed system will not only sound good subjectively, but will objectively meet stringent performance specifications. Currently it is popular to utilize computer modeling to predict the performance of a system in a room. Computer modeling is not the same as a performance specification. A computer model may be a useful sales tool and may provide some valuable engineering data for the systems designer, but at the end of the day it is not how the virtual system in the virtual sanctuary performs. The church must insist that the *installed* system meets a performance specification, or at the very least does what the computer model says it will do.

Five Questions

In the introduction to the chapter on "Sound System Design" in the *Handbook for Sound Engineers*, Chris Foreman indicated that there are five questions that must be answered when designing a sound system:[2]

- Question 1: Is it loud enough?
- Question 2: Can everybody hear?
- Question 3: Can everybody understand?
- Question 4: Will it feed back?
- Question 5: Does it sound good?

Although there are some significant differences in design and function in systems for each of the four worship styles, these five questions apply to all churches regardless of worship style. It is the responsibility of the systems designer to answer each of these questions when considering a design. It is the responsibility of the *church* or its agent to ensure that these questions are answered to the satisfaction of the church.

Question 1: Is it Loud Enough?

This question stems from problems faced by designers working in noisy environments. Don Davis, the founder of Syn-Aud-Con, used to use the example of a racetrack when teaching students about proper engineering in system design. Imagine that the specification was 100 dB SPL in every seat in the stadium. The system was complete and installed and the proud designer was ready to conduct a proof of performance demonstration for the owner. The 25,000 watts worth of power amplifiers were turned on and pink noise was played through the system. The sound levels meters read 97 dB with everything wide open. Don's point was that this 3 dB error would likely mean *doubling* the electrical power! Clearly the question of will it be loud enough is important. In churches we need to ask both *will it be loud enough* and *will it be too loud?* Loud enough for most is making sure that the system produces levels that exceed the background noise (very easy in churches, very difficult in race tracks) and allows each listener in the room to experience speech and music in an appropriate way.

But what about systems that are too loud? The technology exists today to build loudspeaker systems that can, and sometimes do, cause permanent hearing damage to those unfortunate enough to be exposed to them. This is a problem that needs to be addressed mostly in the churches that adopt an experiential worship style, and we will revisit the question in Chapter 22, but it is a question that needs to be addressed repeatedly and aggressively. This is a question with many dimensions, but first people need to be aware that "too loud" in fact does exist, and exists in churches.

Imagine that you are sitting in the pew of your usual church and the pastor gets up and from the pulpit says that he wants everyone to stand up, go outside, lay down in the parking lot, and stare at the sun for 10 minutes. Would you do it? There are a number of ways to approach this question. You could look at it as a question of submission to the authority of the Church. You could also look at it as a matter of having enough faith and trust that God would protect you in your obedience. You could also have a sense of outrage that the pastor would suggest that you do something that would most likely harm you. Ultimately you would have the choice to either comply or not. However, what if the decision was not up to you? What if the pastor picked up a powerful laser and started pointing it into your eyes? This would be outrageous behavior and would most likely result in lawsuits or prosecution. However, something like this scenario is happening in our churches every Sunday across America. Only it is not with light—it is with sound. For some reason we are more likely to understand the dangers of bright lights and lasers to our eyes than we are to understand the dangers of loud sounds to our ears. The dangers are just as real, just as certain, and just as devastatingly permanent. It just might take longer to experience the effects.

There are three well-known standards for noise exposure: the EPA, ANSI/NIOSH, and OSHA have all published standards. According to ANSI/NIOSH, if the sound levels are averaging 98 dB SPL (a weighted, abbreviated dBA) the maximum allowable limit is 23 minutes! If the levels are at 100 dBA the limit is 15 minutes, at 105 dBA the limit is just under 5 minutes! If you turn down to 94 dBA, the limit is 1 hour. The OSHA allows significantly higher doses, however some experts believe that OSHA does not adequately protect workers' hearing. The EPA, which is the most conservative, says that the limit for 94 dBA is only 6 minutes and anything over 100 dBA is considered not safe for any amount of time. Clearly it is the responsibility of the operator to set safe levels. Does the system designer bear any culpability, either legally or morally?

Question 2: Can Everybody Hear?

This question is about coverage and inclusivity. Sound systems need to adequately cover all parts of the sanctuary so that all are able to hear. It also reminds us that the Americans With Disabilities Act mandates that public buildings including churches are required to provide for those with hearing issues.

Question 3: Can Everybody Understand?

Chapter 23 addresses the complex issue of intelligibility more fully. The ability to understand the spoken word is different than the ability to hear. The ability to understand involves aspects of the acoustics of the room as well as attributes of the sound system.

Reverberation (Chapter 21), noise (Chapter 22), and the frequency response of the sound system all have a role in determining if the people in the pews will be able to understand the word spoken from the pulpit. No sound system design or proposal should be considered without an intelligibility specification.

Question 4: Will the System Feedback?

Feedback is that annoying squeal that occasionally happens in sound systems. Most people think that it happens because the sound from the loudspeaker is getting into the microphone. This is partly true. Feedback is the result of nonlinearities in the system. A sound system with perfect components in a perfect room would not feed back. Perfect systems and rooms do not exist obviously, but systems can be designed and built that offer very good stability over a wide range of levels. Although complete freedom from feedback is not something that a reputable sound system designer will guarantee, paying attention to the Potential Acoustic Gain (PAG) and the Feedback Stability Margin (FSM) will ensure that the system is as stable as possible.

POTENTIAL ACOUSTIC GAIN

At the risk of oversimplifying a very complex problem, the potential gain of a sound system depends on a number of elements—the distance between the talker and the microphone (D_s), the distance between the loudspeaker and the microphone (D_1), the distance between the loudspeaker and the furthest listener (D_2), the distance between the talker and the furthest listener (D_0), and the number of open microphones (NOM). Reverberation and reflections also play an important role but are far more difficult to quantify. The equation $PAG = 20\log\dfrac{D_o D_1}{D_s D_2} - 10\log NOM - 6\,dB$ gives a handle on the gain issues in a sound system. The potential gain (PAG) is increased as D_s and D_2, become smaller and as NOM is reduced. The more open microphones are added to a system, the lower the potential acoustic gain. The −6 dB is there as a feedback stability margin (FSM). It has been found that if the system is operated at least 6 dB below the PAG, it will be stable.

Question 5: Does it Sound Good?

Obviously this sort of subjective question is difficult to predict with certainty. It is a necessary question, however, because it is possible to design a sound system that meets all the other criteria, but sounds like a bullhorn! This aesthetic aspect to the system must be addressed and the owner needs to know what can be expected. Asking for a demo of the loudspeakers that are being considered is a good way to begin to address this question, and asking to audition other churches whose systems utilize similar components is another.

Programming

Programming is the term that architects use when they are in the beginning stages of a design. The program is the document that outlines the uses and scope for the project that is being considered. It is generally the responsibility of the client or owner to determine the program, often with the help of the architect. It is an appropriate term to use in the context of sound systems as well. When a church is considering a sound system either for new construction or a replacement for an existing system, there should be a programming phase where the system is essentially defined, not down to the specific make and model to the equipment, but as to its function. In some small churches, this can be a very simple statement, and may not even need to be written down. In large churches this may be a formal document that takes significant time to develop.

There are five steps in the process of developing a program for purchasing a sound system, described as follows.[3]

1. *Establish Goals and Objectives.* Goals and objectives should be determined with all the stakeholders involved. The

stake-holders include but are not limited to those who use microphones—pastors, lay readers, musicians, etc., and instrumentalists, choir directors, etc.

- What should the system do? Voice only? Reinforce Music?
- Size and composition of music group if there is one.
- Is there/will there be an outreach ministry either via broadcast or distribution of recorded services?
- How many rooms (i.e., sanctuary only or is there provision for overflow or feeds to other parts of the complex like nurseries, etc.)?
- Who are the operators? What kind of training and experience do/will they have? This will impact decisions about the complexity of the sound system.
- How will/should the system affect worship?

2. *Gather Relevant Information.* Gather all information relevant to the sound system. This would include information about the building including possible locations for equipment, load limits for structure if equipment is to be hung from the structure, and so on. What are the possible strategies for running cable? Visit other similar churches and inspect their sound systems. Talk with the operators about strengths and weaknesses of their system. Are there local building codes that apply? For example are you required to install an assisted listening system? Identify local (or national if appropriate) experts who can be part of the design team. Identify systems integrators who can install and service the equipment and provide training if needed.

3. *Determine Quantitative Requirements.* Create a bullet point list of what the system must do based on the data collected so far. How many microphone channels? How many instruments? How many sources of prerecorded material? How many feeds to how many spaces?

4. *Determine Qualitative Requirements.* This may be the most important aspect of developing a program. This is the performance specification (spec) that must be met if the system is to do its job. The performance spec also gives the owner some recourse if the performance spec becomes part of the contract with the

system integrator. The qualitative requirements or performance spec should include at minimum the following elements:

- Intelligibility Specification in STI or %ALCONS
- Coverage
- Maximum Sound Pressure Level (SPL) at furthest seat
- Frequency response
- Freedom from hum buzz and rattles
- Immunity from RF interference

5. *Create the Design Program Document.* Summarize all the information just gathered and create a concise document that will become part of the Request for Proposal (RFP) that goes out to vendor/installers.

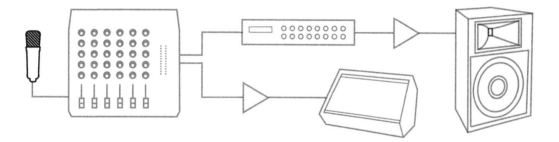

Microphones

Every sound system has the same basic components. We will briefly examine each component and discuss in some detail the choices that are available. At the beginning of the sound system "chain" are the sources of signal, the microphone being the most obvious. Microphones are transducers that transform acoustical energy into electrical energy. There are many ways of classifying microphones. They can be categorized according to the methods of transduction, directionality, usage, diaphragm size, and more. The most important specifications to consider are the frequency response and sensitivity. Microphones are often the least expensive parts of the system and the easiest to change out. It is common to audition or try out different microphones before purchasing to determine the microphone that best suits a specific purpose.

Points to Consider

Generally speaking, wired microphones are better and more cost effective than wireless microphones. Budget wireless systems should be avoided as they will often perform poorly and intermittently. Wireless systems certainly can be used to reduce clutter on the platform and the good ones (read expensive ones) work very well but at a cost sometimes *10 times* the cost of a decent wired microphone. In 2010 the Federal Law changed with regards to which frequencies may be used for wireless microphones. The Buyer should ask the vendor to verify that the wireless microphone meets current FCC regulations.

Get microphones as close to the source of sound as possible. For example, when the source is a voice, it is better to use a miniature boom that puts the microphone close to the mouth rather than a lavaliere mic, which sits down on the chest far away from the mouth.

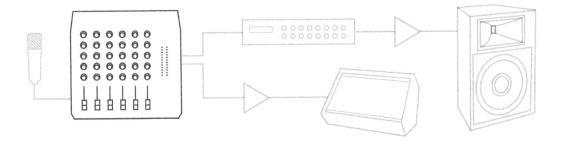

Consoles/Mixers

The console or mixer (also called "board" or "desk") is the device that provides the user control of the sound system. It allows for the combining of microphones and other sound sources and for a certain amount of signal processing as well. It will also control the routing of signals to various destinations as required. Consoles generally are classified according to the number of inputs and outputs. A 16 × 4 console has 16 inputs and 4 main outputs. Another important spec is the number of auxiliary (aux) busses or "sends."

An aux send is an additional output that can be used to send signals other places than the main loudspeakers. Aux sends are used for making recordings, sending signals to stage monitors, sending signals to other areas of the building, and so forth.

Points to Consider

Consoles should be "sized" by considering how many sound sources (microphones, MP3 players, CD players, musical instruments, etc.) are used in a typical worship service. Allow for 25% to 50% more inputs, and as most consoles are built in multiples of 8, round up to the nearest larger size. For example, if a church uses a total of 18 sources every Sunday, a conservative choice would be a 24 input console. This will allow for expansion and for special events.

The complexity of a console should be considered very carefully, taking into account the level of training and experience of the operators. Top of the line consoles offer tremendous flexibility and allow the sound to be very creatively tailored to a given worship service, but can be very intimidating to novice operators.

The question of analog versus digital will still be on the table for a number of years. There may very well come a time when there is no longer a choice, as there will be no more analog choices that are practical. At the time of this writing, there is no clear winner. Digital is not necessarily always better than analog. There are very fine analog consoles, and there are very fine digital consoles. There are also examples of both kinds of consoles that churches should avoid. Analog consoles that are at the very bottom of the price spread should probably be avoided due to concerns over reliability. Digital consoles where the GUI or Graphical User Interface is very idiosyncratic should be avoided by those churches who use many different operators, and the more unusual consoles have the smallest user base.

There are still some who feel strongly that analog consoles sound better than digital. The contribution that the console makes to the overall sound is subtle as compared to loudspeakers and microphones, but it is a legitimate factor to consider. Digital consoles often have a steeper learning curve than their analog counterparts. Because analog consoles are often far less complex, they can be more reliable. The advantage to digital lies mainly in the ability to store preset values and settings and the ability for some of them to be operated in various modes. Some digital consoles can be set up to be operated in a very simple or "novice" mode, where most of the controls are inoperative and only a few necessary ones do anything. This allows for very inexperienced operators to adjust the level of a microphone without having to learn a very complex system. The same console, however, can be used in "expert mode," where the full complexity can be used to good advantage by someone who knows how.

A very important consideration with respect to consoles is the location of the mix position in the sanctuary. There are many churches who apparently feel that the task of mixing is somehow related to broadcasting and relegate the mixing to an isolated booth. This is a very important error! It is virtually impossible to mix a live event like a church service while sitting behind glass. Mixing a live event is not the same as mixing for recording or broadcast. The live operator must be able to hear exactly what the congregation is hearing if they are going to be held responsible for the quality of the sound. In an ideal situation (and where money is no object) each mix of a particular service would be carried out on its own console by people well versed in that particular form of mixing. For example, if a large church has a complex music group leading worship with many musicians and singers, chances are they will need monitors and someone responsible for mixing monitors. Generally the monitor mixer is located directly off stage so there can be visual contact between the musicians and the monitor mix position. The main sound reinforcement system, known in the trade as the front of house (FOH), is mixed from a position somewhere in the sanctuary seating area, by an individual who has experience in FOH mixing. If the church has a broadcast ministry there will need to be a third mixing location isolated from the sanctuary where the production for the broadcast can take place. This is not uncommon in larger churches and requires the services of very competent system integrators to plan, design, and install these complex systems. In smaller churches sound operators are often asked to do double or triple duty, running the front of house mix, the monitor mix, and a recording mix from one console. Most often these operators are untrained and fairly unskilled and are asked to perform a task that even the most seasoned operators may find difficult.

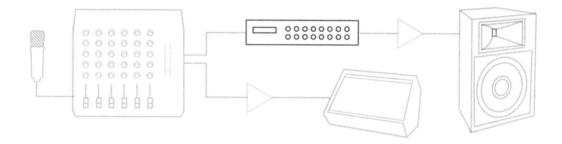

Signal Processors

In the order of signal flow, when the signal leaves the console, it usually is routed to some sort of signal processor. This is one area where digital technology dominates the field. Digital signal processors combine in one unit the functions of many stand-alone analog processors. Crossovers, equalizers, compressor/limiters, delays, and proprietary processing required for some loudspeaker systems all can be handled by a single digital processor. In addition, many offer automatic feedback suppression. These suppressors work reasonably well if set up correctly, but cannot guarantee that the system will never go into feedback. There are not many pitfalls in choosing a signal processor. They are all very similar in how they behave, and it is not difficult to find a unit that fits the technical requirements while staying within budget. Some digital signal processors have the advantage of no user-accessible controls. In churches where there are operators who are not professional and have little experience, it is a good idea not to let the operators get access to the digital signal processor. The DSP will be set up by the installer and if it is done correctly, the adjustments will rarely need to be changed.

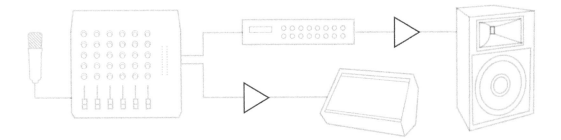

Amplifiers

Next in the signal chain are the power amplifiers that amplify the signal to a high enough level so the loudspeakers can then turn that power into sound. The choice of amplifier should be made on the basis of reputation for ruggedness and overall sound. Like the console, the amplifier will contribute to the overall sound of the system and a sound mixer is not necessarily pulling a fast one when he insists on a particular brand of amplifier because of its sound! Probably the most common mistake that is made when selecting power amplifiers is choosing amplifiers that are too small or underpowered for the job. For a variety of reasons, amplifiers that are too small (i.e., they can't deliver enough power) are more likely to cause loudspeaker failure than amplifiers that are too large. Amplifiers generally produce a fair amount of heat and also can be mechanically noisy due to internal fans. This may not be a problem in small systems with one amplifier but in large systems with many amplifiers, the heat, noise, and power consumption are all significant design factors.

Loudspeaker Cable

It may seem odd to dedicate a portion of this chapter to wire. Wire and "best–practice" wiring and termination techniques are important at all the stages of a system to insure trouble-free operation and freedom from external interference. However, the wire between the power amplifier and the loudspeaker is a special case. This is the only wire in the system other than the power cords, which connect the equipment to the mains, that is responsible for *power* transfer. In all other cases the wire is responsible for *voltage* transfer. Voltage, as potential energy, can travel long distances through wire with negligible losses. Power (watts) is current (amperes, I) times voltage (E) or $P = I \times E$. This is no longer potential energy but kinetic. The loudspeaker is moving compressing the air! Work is being done. Without getting too far into the theory, the wire matters! There are numerous examples of large sound systems where fully half of the power delivered by the power amplifier is dissipated not by the intended loudspeaker but by the wire connecting the loudspeaker to the amplifier! This is just bad engineering and should be avoided. In general the wire that connects the loudspeaker to the amplifier needs to be large, a minimum of 12 gauge, and the runs need to be as short as possible.

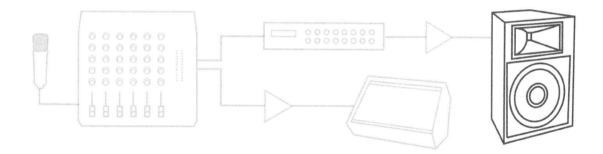

Loudspeakers

Loudspeakers are the most important part of a sound system in the sense that they are the elements that will be most influential in determining how a system will sound. Loudspeakers are the weakest links in the chain, mostly because they are the most difficult part of a sound system to build. Loudspeaker systems are challenging from an engineering point of view due to the wide bandwidth required. The lowest frequency sound we can hear and therefore need reproduce is 20 Hz with a wavelength of over 50 feet. Theoretically, average humans can hear as high as 20,000 Hz (20 kHz) with a corresponding wavelength of 0.6 inches. Even if we allow that most of us cannot hear much above 10,000 Hz (1.2 inch wavelength) we are still are left with an enormous range of wavelengths that we need to reproduce faithfully from a single device. A person with average hearing has a *dynamic range* (the range of quietest sound audible to the loudest sound before hearing damage) of around 120 dB. That is a ratio of a trillion to one! Fortunately most music in the average church service will probably have a dynamic range of only 50 dB or less, so we are not asking loudspeakers to be able to reproduce the entire range of loudness. To make matters even more complicated, due to the physics of loudspeaker systems, it is fairly difficult to get sound to go where you want it to go. Low frequency loudspeakers tend to be omnidirectional; that is, the sound they create goes everywhere. High frequency loudspeakers tend to create narrow beams. Neither one of these extremes is ideal. An ideal loudspeaker would cover just the seating area, equally well at all frequencies, and exhibit the same response at all seats. All loudspeaker systems are compromises, and each of the compromises is audible.

We will consider first loudspeakers in terms of function (e.g., monitors vs FOH), then with respect to implementation (e.g., central cluster vs distributed).

Function

Monitors

All loudspeakers turn electrical energy into acoustic energy, but there are many differences in application. Monitors are loudspeakers that have been optimized for a specific application. The term monitor as it applies to live sound reinforcement refers to a loudspeaker that is used by performers on stage to allow them to hear themselves and the other musicians as well. The first question that must be asked is, are monitors necessary? Most musicians will immediately answer, of course, as if it is, or should be, obvious! Monitors are not a panacea. Adding monitors to the stage requires a level of sophistication and understanding that many church sound operators simply do not have. If the church is small enough and the musicians can find away to hear themselves without monitors this is always a better strategy.

Wedges

Assuming that it is decided that monitors are necessary, there are basically three types of monitor systems that can be considered. The first and most common type is the wedge. These are loudspeakers designed to sit on the stage generally in front or upstage from the microphone stand, and point at the performer. Whenever loudspeakers are placed close to live microphones it requires a considerable amount of skill on the part of the operator to keep the systems from feeding back. The advantage to this type of system is that it is flexible and often many musicians can share a monitor. The downside is that in situations where the band is quite loud, for example when there is a live drummer, the monitors will have to be quite loud to allow the musicians to hear themselves above the loud drums and other amplified instruments on stage. Loud levels from stage monitors will often bleed into the audience area (the house) and cause problems for the FOH mixer.

Hot Spot

A second type of monitor knows as a hot spot can be used. The hot spot is a small loudspeaker attached to a microphone stand, and can be placed close to the musicians' ears so that it does not have to be turned up as loud as a wedge might need to be. The main drawback is that they tend to add clutter to already crowded stages.

In-Ear

In recent years there has been a trend toward using in-ear monitors. They have the huge advantage of reducing the levels on stage and eliminate bleed from monitors into the house. The main disadvantage to in-ear monitors is the cost. In-ear monitors are usually custom-made for each musician. The custom-made earpieces are connected wirelessly to the console providing the monitor mix.

FOH Loudspeakers

There are a huge variety of loudspeakers and loudspeaker systems available for FOH use. There are loudspeakers that will fit every budget and size of venue. As mentioned earlier, the ideal loudspeaker would handle the full range of frequencies, and cover only the seating area, not sending any sound into the rest of the room where it does no good and in fact can be problematic. Earlier in this chapter we referred to qualitative requirement, which are objective performance specifications. The degree to which the system will meet these specifications is largely due to the choice of speaker type.

Direct Radiators

Broadly speaking there are three categories of loudspeakers used for FOH application: direct radiators, horn-loaded loudspeakers, and line sources or arrays. Direct radiators are loudspeakers with little or no control over where the sound goes. These are the most inexpensive and may be fine for smaller churches, especially churches who worship in the Evangelical and Community styles. They should be avoided by larger reverberant churches like the Celebratory, unless part of a carefully designed system. Sometimes high frequency loudspeakers are connected to horns whereas low frequency loudspeakers are direct radiating.

Horns

Horns are devices that direct the sound of a loudspeaker to the areas where it needs to go—for example, the audience of a church—and minimizes sending sound to where it does no one any good—like up into a ceiling vault. Horns are identified by three descriptors: the frequency range that the horn will control, the horizontal and vertical coverage angles, and the directivity index sometimes referred to as Q. In order for horns to control low frequencies, they need to be physically large. If a loudspeaker system claims to be fully horn-loaded but the loudspeaker cabinet is physically small, something is not right. Coverage angles are by convention expressed by listing the horizontal coverage first, then the vertical. A 6040 horn covers a 60 degree arc horizontally and a 40 degree arc vertically.

An important measure of the effectiveness of a horn is the directivity index (Q). The DI or Q is a function of the horn design but is also a result of the desired coverage area. A 90-degree–wide horn should have a lower Q than a 40 degree horn so a higher number is not always better. What should be examined very carefully is a graph that plots the directivity as a function of frequency. A good horn will maintain control over a broad range of frequencies.

Figure 20.1 Altec Lansing Manta-Ray Horn Family; Cast Aluminum Throat and Soldered; Coated Bell Construction.

Courtesy Altec Lansing Corp.

Figure 20.2 JBL Biradial Horn Family; Cast Aluminum Throat; Fiberglass Bell Construction.
Courtesy JBL/UREI.

Figure 20.3 Electro Voice HR 9040 Constant Directivity Horn.

In the early days of audio, horns ruled. Horns made it possible to improve the efficiency of a loudspeaker system dramatically, because instead of a loudspeaker filling a room with sound as a floodlight might fill a room with light, a horn can focus the energy where it is needed, much like a spotlight might do. This was very important when power amplification was expensive and difficult. Some of these great horns of the past are shown in figures 20.1, 20.2 and 20.3. As technology moved on, audio people began to realize that if you were willing to sacrifice efficiency, direct radiators sounded better. Many were willing to accept the loss of control and efficiency for better sound. Horns began to get a bad reputation. We still see audio sales reps for companies who don't build horns cupping their mouths with their hands and saying, "We don't use horns because they sound bad... don't you hear that honking sound?" A fuller understanding of the physics of how horns work has resulted in horns that are considerably more complex than cupped hands and exhibit well-controlled directivity that remains constant over frequency (constant directivity) as well as devices that sound as good as direct radiators (and in some cases, better). A number of companies, most notably Community and Danley Sound Labs continue to build very high quality great sounding horns.

Figure 20.4 Danely SH50, Full Range Horn.

Line Sources

The third loudspeaker type is the line source or line array. The other two loudspeaker types are based on a point-source model where the actual source of the sound may be thought of as a point in space radiating spherically or omnidirectionally. The horn then tries to constrain all the energy from the theoretical point into a coverage pattern determined by the horizontal and vertical coverage angles. The line source is based on a totally different model. A theoretical point source would be a dimensionless point in space that would radiate sound in all directions; a theoretical line source would be a line of infinite length that would radiate not as a sphere, but rather as a cylinder. The main advantage of a line source is that if one could be built, the energy would dissipate much more slowly. Without getting too technical, a point source radiating spherically will lose 6 dB of sound pressure (loudness) every time the distance away from the source is doubled, but a line source will lose half that, only 3 dB, for every doubling of distance. In the real world, line sources can only be approximated. One way is to build very long columns of very small loudspeakers. Loudspeaker systems like the IntelliVox® by Duran Audio, the Encases® by Community, and the ICONYX® by Renkus-Heinz are all examples of line sources built from many small loudspeakers. Some of these systems, like the IntelliVox and the ICONYX, are electronically steerable, allowing the system to be carefully matched to the needs of the room. Theses systems are generally very good as speech systems in very reverberant spaces. These systems do not offer much, if any, control over the horizontal coverage and they are generally over 100 degrees wide, and vary considerably with frequency.

The line array is an attempt at building a line source out of individual broadband loudspeaker *cabinets*. (Figure 20.5) The line array has become very popular in recent times and is aggressively marketed by the manufacturers. They tend to be very expensive and require extensive rigging to hold them in place. In spite of the popularity of the line array there are a growing number who question the huge cost and question the effectiveness of the array. One of the severe drawbacks of the line array, even a theoretically perfect one, is that the frequency response will be different virtually everywhere in the coverage pattern that you measure it.

Line arrays are approximations of a true line source, and as such their response can and does vary considerably over the coverage. The church is well advised to insist that if a line array is used, the performance specifications, especially those dealing with coverage, frequency response and intelligibility are met.

Implementation

It is important for a church to decide on the type of loudspeaker. It is equally important to choose the location or locations of the loudspeakers. The exact position of the loudspeaker(s) should be determined by a professional, but there are decisions in which the church should be involved. A governing principle is that when it comes to loudspeakers, less is more. A good analogy is to

Figure 20.5 A Typical Line Array. Courtesy Yorkville.

consider being an observer in a church where there is responsive reading. During a responsive reading the minister reads a line, then the congregation in unison reads a line. The single voice alone is always much easier to understand then the group. If many loudspeakers are used in a room, great care must be taken to prevent the sound from more than one loudspeaker reaching any particular listener. The best location for a single loudspeaker or loudspeaker system is as near the true source of the sound as possible. It is beyond the scope of this chapter to go into all the details of why this is important. For a small evangelical style church the best place for a loudspeaker system is above and in front of the pulpit. In this orientation each member of the congregation is getting the direct sound from only one source and the source is physically near the talker. A central location is always better than placing two loudspeaker systems on either side of the stage.

Sometimes it is just not possible to cover a room from one location. This can happen when there is a balcony with seating underneath and the under balcony seats can't see the main cluster. It can also happen when rooms are long—that is, the furthest seat is more than 75 to 80 feet from the stage—additional loudspeakers or clusters of loudspeakers must be used. These auxiliary loudspeakers are often referred to as delay loudspeakers or clusters. Since sound travels relatively slowly through the air (approx. 331 meters per second or 1128 feet per second), these loudspeakers must be electronically delayed to synchronize the sound emanating from the remote clusters with the primary cluster. This synchronization is absolutely essential if intelligibility is to be achieved at all. If the delays are not used, or are misadjusted, the people seated in the areas covered by the auxiliary cluster(s) will likely hear the sound from many sources, and intelligibility will be seriously impaired. It should be noted that when the delays are set correctly, the delayed cluster is not perceived as the source of the sound. It is not uncommon for people to complain to the sound operator that the speakers that are the closest to them are not operating!

Another variable in the choice of loudspeaker implementation is the question of a stereo versus monaural system. Generally speaking monaural systems are best for speech reinforcement.

Stereo systems are difficult to implement well in large rooms as relatively few people, and only those seated in the middle of the room, will experience the stereophonic effect. A common implementation for those churches that use prerecorded tracks as musical accompaniment to a choir, for example, is to have a center loudspeaker or cluster responsible for speech and live music reinforcement and a stereo system consisting of loudspeakers on either side of the stage for playback of tracks. These systems are known as L-C-R or left, center, right systems, and can be effective in the hands of a skilled sound mixer.

Subwoofers

Subwoofers (subs) are generally large loudspeakers that cover only the very lowest bass notes. They can add significant impact to the music, especially if bass guitars and electronic keyboards are used in worship. Many churches, especially those who utilize the experiential form of worship, feel that subwoofers are indispensable, and for that style of worship maybe they are. However, it can be very difficult to make subs work well and there are many churches that have spent considerable amounts of money on subs only to find that they can't use them. Smaller churches, especially those who worship in the evangelical and community styles, should very carefully consider the use of subs. They represent a significant investment for less than an octave of musical notes. Most subs are omnidirectional. It takes some high tech solutions to make bass go where you want it to and not everywhere in the room. For this reason subs in highly reverberant spaces are generally not a good idea.

End Notes

1. Carolyn, D., & Don, D., (2004). *If Bad Sound Were Fatal, Audio Would Be The Leading cause of Death,* 1stBooks Publisher, Bloomington IN
2. Ballou, G. (Ed.), (2008). *Handbook for Sound Engineers* (4th ed., p. 1239). Focal Press.
3. Adapted from Whole Building Design Guide, http://www.wbdg.org/design/dd_archprogramming.php, accessed 1/10.

REVERBERATION AND TIME RESPONSE

Reverberation

In 1898, Wallace Sabine, a physicist at Harvard University, developed a formula that defined the relationships between the volume of a room (V), the total surface area (S), the absorption coefficient of the materials in the room, (α), and the time it will take for a sound to decay to one millionth of its original intensity RT_{60}. The formula $RT_{60} = \dfrac{.161V}{S\alpha}$ (use 0.049 as constant for feet) that he developed purely by observation is still used today. Over the years there have been developed a few different forms of the basic equation that may yield more accurate results in certain conditions. In his paper describing this equation he includes a comment that is remarkable in its prophetic nature. In a section of the paper dealing with terminology he addresses the difference between resonance and reverberation. He says, "In scientific literature the term resonance has received a very definite and precise application to the phenomenon, where ever it may occur, of the growth of a

Sound of Worship. DOI: 10.1016/B978-0-240-81339-4.00021-1

vibratory motion of an elastic body under periodic forces timed to its natural rates of vibration. A word having this significance is necessary; and it is very desirable that the term should not, even popularly, by meaning many things, cease to mean anything thing exactly."[1]

This is precisely what has happened to the term reverberation. By meaning many things, it has ceased to mean anything at all. Reverberation, at least in popular usage, means anything from a guitar effects pedal to some ill-defined sense of the sound of a space. In room acoustics it has become a simplistic, one number descriptor of a room, with the implication that if you know the "reverb time," you have a complete description of the acoustics. Nothing could be further from the truth.

Before we examine reverberation, it might be useful to review some basic acoustics. Sound is made up of vibrations traveling in the form of a longitudinal wave through a medium, usually air. It will continue to travel until the energy is dissipated in the medium or until it impinges on an object. The size and composition of the object will determine what happens to the impinging wave. If the object is small relative to the wavelength or size of the sound wave, and reflective, the wave will diffract around it and continue as if the object was not there (Figure 21.1).

If the object is large relative to the wavelength and reflective, the wave will be reflected back. The direction that the reflected wave will take is determined by the angle of incidence of the wave; the angle of incidence will be equal to the angle of reflection.

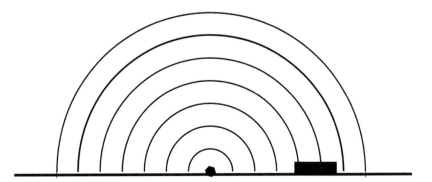

Figure 21.1 Diffraction.

For material to be reflective, it is usually nonporous and rigid, and most of the acoustical energy impinging on it will be reflected (Figure 21.2).

If the material is smooth, that is to say the irregularities in the surface are very small relative to the wavelength of the impinging wave, the reflection will be specular. Irregularities in the surface will cause the energy to be diffused or reflected back in non-specular directions. Material is considered to be diffusive if the impinging wave is reflected back and scattered in many, if not all, directions.

Material is considered absorptive if it has some property that can convert the acoustical energy into some other form of energy, usually heat. Absorptive materials are usually porous, soft, or nonrigid.

Every real-world material exhibits some combination of each of these three properties. Most material will exhibit different behaviors depending on the wavelength of the impinging wave. Consider a large pane glass window. At very low frequencies (large wavelengths) the window might be considered an absorber as there *may* be enough energy in the low frequencies to make the glass bend and act as a diaphragm, the acoustical energy being transferred into mechanical energy and ultimately heat through frictional losses in the glass. At higher frequencies (shorter wavelengths) the window would act as a very effective reflector. At very high frequencies in the ultrasound range (very short wavelengths) the glass may begin to behave as a diffuser

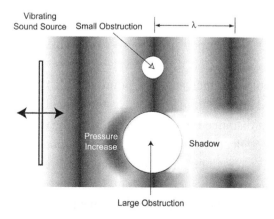

Figure 21.2 **Sound Interacting with Surfaces.**

as there may be enough tiny irregularities in the glass that are of the same order of magnitude as the ultrasonic wave hitting it so scattering may occur. This is obviously an extreme example, but it should be remembered that all building materials and furnishings will affect the way sound behaves and is ultimately perceived in a space, some in dramatic ways, some in very subtle ways.

The dictionary defines reverberation as "the persistence of a sound after its source has stopped." That is a good start, but according to Sabine's original paper, there are two criteria that must be met if true reverberation is to exist.

The first criterion is that the sound field be homogeneous. In a true reverberant space, no matter where we sample the sound field we will get the same result. In layman's terms, walk anywhere in a truly reverberant space and the quality and volume or amplitude of the sound will not change. To illustrate this, imagine two people in a large cathedral. If they stood a few feet apart, a conversation could occur easily. If the two started to move apart, at some point it would be no longer possible to converse. The voices would be *audible* but they would not be *intelligible*. This would be close to the theoretical critical distance. The following graph represents the level in decibels vs distance in feet from a theoretical point-source in a true reverberant space. Initially the level drops as the distance increases according to the inverse square law, where the level will drop 6 dB for every doubling of distance. The dashed line represents what happens in free field, or in a nonreverberant space. In free space, the sound level will continue to drop until it is totally dedicated or it reaches a level equal to the noise floor. The solid line depicts what will happen in a reverberant space. As the distance between the source and the observer increases, the contribution from the room will increase. The point at which the dashed line and the solid line separate is known as the critical distance (Figure 21.3). At that distance the contribution from the direct sound and from all the reflected sound is equal. As distance increases in a reverberant space, the contribution from the direct sound diminishes, but the reflected energy stays on for some finite time and, as long as the source stays on, the level is constant or close to it.

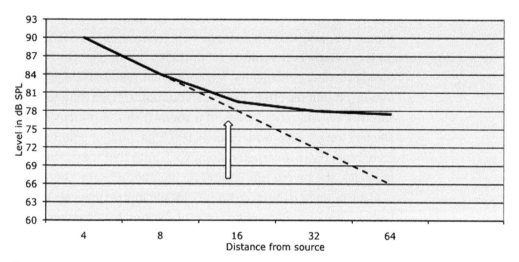

Figure 21.3 Critical Distance.

If we were to try the same experiment in a room that is acoustically rather dead or nonreverberant, we would find that there is no distance where the reflected energy and the direct energy are equal. As we move away from the source, it just gets quieter, but the reflected sound field never develops and takes over. If there is no critical distance, it is unlikely that there will be a true reverberant field.

The second criterion for reverberation to occur is isotropy. Isotropy is the property of the sound field where the sound energy is equally likely to flow in any direction, or the sound energy flows equally in all directions. If we can tell where the sound is coming from, this is not an isotropic condition and therefore it is not a reverberant field.

We might ask, so what? There are a number of equations[2] that include reverb time as a variable. If there is no true reverberation present in a particular room, the equations that might be used to predict other aspects of the space based partially on reverb time will not work.

Reverberation is frequently misused and misunderstood. Common mistakes include referring to a reverb time as one number. We are all guilty of this one—even acoustical consultants who should know better occasionally refer to a room as a "2.5 second room," meaning that the room has a reverberation time of 2.5 seconds generally in the mid-band or mid-range frequencies. In our basic overview of acoustics, we noted that materials behave differently at different frequencies. So do rooms. If we were to measure the reverberation times of a room with the dimensions of 100 feet long, 75 feet wide, and 40 feet high, built of painted concrete block, with a carpeted floor, we would find that at low frequencies, the reverb time would be high because the coefficient of absorption of painted concrete block at low frequencies is negligible. If we examine Sabine's formula written in its modern form $RT_{60} = \frac{.049V}{S\alpha}$, we see that the variable α is a very powerful variable. The variable α or the coefficient of absorption is expressed in Sabine units (appropriately named after W. C. Sabine) and the values range from 0 indicating total reflectivity, to 1.0 indicating total absorption. Sabine defined a coefficient of 1.0 as the absorption of 1 square foot of open window. A more useful form of the equation, the Norris Eyring equation, $RT_{60} = \frac{.161V}{S_1\alpha_1 + S_2\alpha_2 + S_3\alpha_3 + \cdots S_n a_n}$ is where we can individually calculate the effect of each surface and the enter the absorption coefficient associated of the material on each surface. There are many databases where we can find the absorption coefficients for various building materials. Figure 21.4 provides a short excerpt for illustration sake.

According to the table in Figure 21.4, at 125 Hz, a relatively low frequency, painted block has a coefficient of 0.1 and carpet has a coefficient of 0.02! At this frequency the concrete block actually absorbs more than the carpet does! If we plug these numbers into the modified sabine formula we get a reverberation time of almost 8 seconds at 125 Hz. If we look at a higher frequency of 2000 Hz, the table tells us that the block still has an absorption coefficient of 0.1 but the carpet has a coefficient of 0.6. The reverberation time at 2000 Hz is 2.3 seconds. If we graph this room using the data from the table, it would be graph 2, as shown in Figure 21.5.

What number would accurately describe the reverberation time of this room?

Materials	Coefficients					
	125 Hz	**250 Hz**	**500 Hz**	**1 kHz**	**2 kHz**	**4 kHz**
Acoustical plaster ("Zonolite")						
½ in. thick trowel application	0.31	0.32	0.52	0.81	0.88	0.84
1 in. thick trowel application	0.25	0.45	0.78	0.92	0.89	0.87
Acoustile, surface glazed and perforated structural clay tile, perforate surface backed with 4 in. glass fiber blanket of 1 lb/ft^2 density	0.26	0.57	0.63	0.96	0.44	0.56
Air (Sabins per 1000 ft^3)					2.3	7.2
Brick, unglazed	0.03	0.03	0.03	0.04	0.05	0.07
Brick, unglazed, painted	0.01	0.01	0.02	0.02	0.02	0.03
Carpet, heavy						
on concrete	0.02	0.06	0.14	0.37	0.60	0.65
on 40 oz hairfelt or foam rubber with impermeable latex backing	0.08	0.24	0.57	0.69	0.71	0.73
on 40 oz hairfelt or foam rubber						
40 oz hairfelt or foam rubber	0.08	0.27	0.39	0.34	0.48	0.63
Concrete block						
coarse	0.36	0.44	0.31	0.29	0.39	0.25
painted	0.10	0.05	0.06	0.07	0.09	0.08
Fabrics						
light velour, 10 oz/yd^2, hung straight in contact with wall	0.03	0.04	0.11	0.17	0.24	0.35
medium velour, 10 oz/yd^2, draped to half area	0.07	0.31	0.49	0.75	0.70	0.60
heavy velour, 18 oz/s yd^2 draped to half area	0.14	0.35	0.55	0.72	0.70	0.65
Fiberboards, ½ in. normal soft, mounted against solid backing						
unpainted	0.05	0.10	0.15	0.25	0.30	0.3
some painted	0.05	0.10	0.10	0.10	0.10	0.15
Fiberboards, ½ in. normal soft, mounted over 1 in. air space						
unpainted	0.30		0.15		0.10	
some painted	0.30		0.15		0.10	
Fiberglass insulation blankets						
AF100, 1 in., mounting # 4	0.07	0.23	0.42	0.77	0.73	0.70
AF100, 2 in., mounting # 4	0.19	0.51	0.79	0.92	0.82	0.78
AF530, 1 in., mounting # 4	0.09	0.25	0.60	0.81	0.75	0.74
AF530, 2 in., mounting # 4	0.20	0.56	0.89	0.93	0.84	0.80
AF530, 4 in., mounting # 4	0.39	0.91	0.99	0.98	0.93	0.88
Flexboard, 3/16 in. unperforated cement asbestos board mounted over 2 in. air space	0.18	0.11	0.09	0.07	0.03	0.03
Floors						
concrete or terrazzo	0.01	0.01	0.015	0.02	0.02	0.02
linoleum, asphalt, rubber, or cork tile on concrete	0.02	0.03	0.03	0.03	0.03	0.02
wood	0.15	0.11	0.10	0.07	0.06	0.07
wood parquet in asphalt on concrete	0.04	0.04	0.07	0.06	0.06	0.07

Figure 21.4 Absorption Coefficients.

Materials	Coefficients					
	125 Hz	**250 Hz**	**500 Hz**	**1 kHz**	**2 kHz**	**4 kHz**
Geoacoustic, 13½ in. × 13½ in., 2 in. thick cellular glass tile installed 32 in. o.c. per unit	0.13	0.74	2.35	2.53	2.03	1.73
Glass						
large panes of heavy plate glass	0.18	0.06	0.04	0.03	0.02	0.02
ordinary window glass	0.35	0.25	0.18	0.12	0.07	0.04
Gypsum board, ½ in. nailed to 2 in. × 4 in., 16 in. o.c.	0.29	0.10	0.05	0.04	0.07	0.09
Hardboard panel, ⅛ in., 1 lb/ft^2 with bituminous roofing felt stuck to back, mounted over 2 in. air space	0.90	0.45	0.25	0.15	0.10	0.10
Marble or glazed tile	0.01	0.01	0.01	0.01	0.02	0.02
Masonite, ½ in., mounted over 1 in. air space	0.12	0.28	0.19	0.18	0.19	0.15
Mineral or glass wool blanket, 1 in., 5–15 lb/ft^2 density mounted against solid backing						
covered with open weave fabric	0.15	0.35	0.70	0.85	0.90	0.90
covered with 5% perforated hardboard	0.10	0.35	0.85	0.85	0.35	0.15
covered with 10% perforated or 20% slotted hardboard	0.15	0.30	0.75	0.85	0.75	0.40
Mineral or glass wool blanket, 2 in., 5–15 lb/ft^2 density mounted over 1 in. air space						
covered with open weave fabric	0.35	0.70	0.90	0.90	0.95	0.90
covered with 10% perforated or 20% slotted hardboard	0.40	0.80	0.90	0.85	0.75	
Openings						
stage, depending on furnishings			0.25–0.75			
deep balcony, upholstered seats			0.50–1.00			
grills, ventilating			0.15–0.50			
Plaster, gypsum or lime						
smooth finish, on tile or brick	0.013	0.015	0.02	0.03	0.04	0.05
rough finish on lath	0.02	0.03	0.04	0.05	0.04	0.03
smooth finish on lath	0.02	0.02	0.03	0.04	0.04	0.03
Plywood panels						
2 in., glued to 2½ in. thick plaster wall on metal lath	0.05		0.05		0.02	
¼ in., mounted over 3 in. air space, with 1 in. glassfiber batts right behind the panel	0.60	0.30	0.10	0.09	0.09	0.09
⅜ in.	0.28	0.22	0.17	0.09	0.10	0.11
Rockwool blanket, 2 in. thick batt (Semi-Thik)						
mounted against solid backing	0.34	0.52	0.94	0.83	0.81	0.69
mounted over 1 in. air space	0.36	0.62	0.99	0.92	0.92	0.86
mounted over 2 in. air space	0.31	0.70	0.99	0.98	0.92	0.84

Figure 21.4 (Continued)

Materials	Coefficients					
	125 Hz	**250 Hz**	**500 Hz**	**1 kHz**	**2 kHz**	**4 kHz**
Rockwool blanket, 2 in. thick batt (Semi-Thik), covered with $3/16$ in. thick perforated cement-asbestos board (Transite), 11% open area						
mounted against solid backing	0.23	0.53	0.99	0.91	0.62	0.84
mounted over 1 in. air space	0.39	0.77	0.99	0.83	0.58	0.50
mounted over 2 in. air space	0.39	0.67	0.99	0.92	0.58	0.48
Rockwall blanket, 4 in. thick batt (Full-Thik)						
mounted against solid backing	0.28	0.59	0.88	0.88	0.88	0.72
mounted over 1 in. air space	0.41	0.81	0.99	0.99	0.92	0.83
mounted over 2 in. air space	0.52	0.89	0.99	0.98	0.94	0.86
Rockwool blanket, 4 in. thick batt (Full-Thik), covered with $3/16$ in. thick perforated cement-asbestos board (Transite), 11% open area						
mounted against solid backing	0.50	0.88	0.99	0.75	0.56	0.45
mounted over 1 in. air space	0.44	0.88	0.99	0.88	0.70	0.30
mounted over 2 in. air space	0.62	0.89	0.99	0.92	0.70	0.58
Roofing felt, bituminous, two layers, 0.8 lb/ft^2, mounted over 10 in. air space	0.50	0.30	0.20	0.10	0.10	0.10
Spincoustic blanket						
1 in., mounted against solid backing	0.13	0.38	0.79	0.92	0.83	0.76
2 in., mounted against solid backing	0.45	0.77	0.99	0.99	0.91	0.78
Spincoustic blanket, 2 in., covered with $3/16$ in. perforated cement-asbestos board (Transite), 11% open area	0.25	0.80	0.99	0.93	0.72	0.58
Sprayed "Limpet" asbestos						
$3/4$ in., 1 coat, unpainted on solid backing	0.08	0.19	0.70	0.89	0.95	0.85
1 in., 1 coat, unpainted on solid backing	0.30	0.42	0.74	0.96	0.95	0.96
$3/4$ in., 1 coat, unpainted on metal lath	0.41	0.88	0.90	0.88	0.91	0.81
Transite, $3/16$ in. perforated, cement-asbestos board, 11% open area						
mounted against solid backing	0.01	0.02	0.02	0.05	0.03	0.08
mounted over 1 in. air space	0.02	0.05	0.06	0.16	0.19	0.12
mounted over 2 in. air space	0.02	0.03	0.12	0.27	0.06	0.09
mounted over 4 in. air space	0.02	0.05	0.17	0.17	0.11	0.17
paper-backed board, mounted over 4 in. air space	0.34	0.57	0.77	0.79	0.43	0.45
Water surface, as in a swimming pool	0.008	0.008	0.013	0.015	0.02	0.025
Wood paneling, $3/8$ in. to $1/2$ in. thick, mounted over 2 in. to 4 in. air space	0.30	0.25	0.20	0.17	0.15	0.10

Figure 21.4 (Continued)

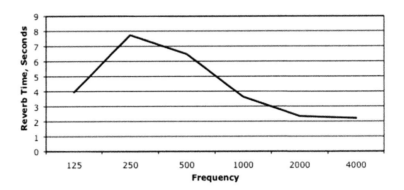

Figure 21.5 Reverb Time vs Frequency of Hypothetical Room.

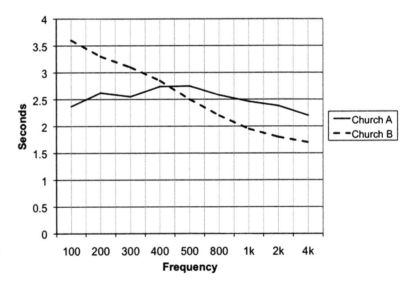

Figure 21.6 Reverb Time vs Frequency of Two Hypothetical Churches.

Suggested Listening at www.sound-of-worship.com

Go to the Reverberation Examples section to listen to the church demo.

Figure 21.6 shows reverberation time versus frequency for two churches. Both have average reverberation times of very close to 2.5 seconds but these two rooms sound remarkably different. One church, church B, sounds rich and full whereas church A sounds thin and harsh. Clearly, one number is not sufficient to describe these rooms.

When designing a church, we must look at the impact of the materials application on the reverberation curve, not simply the average reverb time. For example, choosing between metal and wood studs will change the diaphragmatic behavior of the walls thereby changing how much low frequency energy is absorbed.

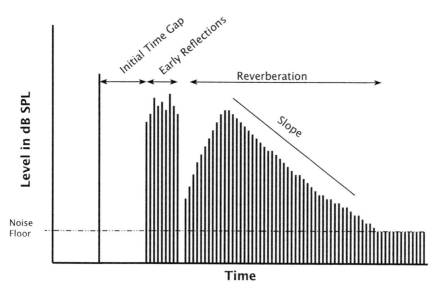

Figure 21.7 Regions in the Time Response of a Large Room.

The acoustic performance of building materials will always be frequency dependent. The choice of studs for example will impact the low frequency performance but at frequencies above approximately 800 Hz the choice of studs will have no impact at all. The same is true for stud spacing. We might choose 24 inch spacing over 16 inch if we have the structural leeway to do so, but 24 inch spacing again makes for much more low frequency absorption than does 16 inch spacing as the walls are significantly more diaphragmatic.

As noted in Chapter 19, the relationship between what we measure or calculate and what we perceive is always complex, especially in acoustics. Figure 21.7 is a representation of an envelope time curve (ETC; also known as Energy Time Curve). The ETC is a measurement of the envelope of the energy over time, or how the amplitude of the sound changes over time. The spike at the left of the graph is the direct sound. The direct sound is the sound that leaves the source and travels directly to the observer with out impinging on any surface. Then there is a gap known as the initial time gap (ITG). This gap represents the time that it takes for the sound to bounce off the nearest surfaces and arrive at the listener. The size or length of this gap corresponds to the physical size of the room. Large rooms have longer initial time gaps as the reflecting surfaces are further away. The amplitude of the early reflections that occur after the ITG play a big role in the perception of how "live" a room feels. Although rooms with long reverb times often sound "live," as it turns out, the primary contributor to the subjective impression of "liveness" is not reverberation time. It is what happens in the first 30 or so milliseconds (ms) after the direct sound that determines the sensation of liveness. If there are many strong reflections in the first 30 ms, the room is judged live. If the reflections are week or sparse, the room is judged dead, even if there are later reflections that comprise a reverberant field. This is particularly important to churches, especially those that value congregational singing very highly. The sanctuary must have a live feel if the congregation is to feel like singing. The reverberant field actually takes some time to develop in a room so Figure 21.7 shows the build-up of the reverberant field rising to a peak, then the decay of the reverberation. It is the slope of the decay that yields the reverberation time.

Time Response

Fortunately there are other parameters that we can and must consider beyond reverberation time. As of this writing, there is no extensive research correlating acoustic parameters to the quality of worship. This sort of study could be crafted and should be. In the absence of good studies, we can make a case for applying well-accepted acoustic principles that have been thoroughly studied and implemented in other contexts. Men like Leo Beranek and Yoichi Ando were pioneers in concert hall acoustics, and worked very hard to connect the subjective experience of listening to a concert to the objective measures of a space.

These men, and others, have shown that the subjective performance of a concert hall can be predicted by detailed analysis of the time response of a room. By comparing the amount of energy present in specific blocs or regions of time we can predict subjective aspects of the space.

Many of the parameters that they identified are now included in ISO 3382, the international standard on room acoustics. ISO 3382 has been accepted as the standard for measuring and describing concert halls and other performance spaces. The goal of a concert hall is to project the sound of the orchestra to every seat in the house such that every member of the audience has an optimal experience of the music. Orchestra halls are generally one-way rooms, that is to say, the acoustics are designed to work in one direction. The sound leaves the stage and travels to the audience. Churches are not necessarily one-way rooms nor are they necessarily performance spaces and some of the criteria set forth in ISO 3382 are probably not relevant to the worship experience. We will examine a few of the criteria of ISO 3382 that we believe are important for churches, and these will be the criteria we use in the chapters dealing with the acoustics requirements of churches employing each of the worship styles.

Reverberance (EDT)

In the past 25 years or so it has become clear that the parameter, which Wallace Sabine so eloquently described in the late 1800s, does not adequately predict the subjective experience of a space. It is possible to have a room that feels acoustical "dry" or "dead" while having a long reverberation time or RT_{60}. It is also possible to have a room that feels acoustically "wet" or "live" but has a fairly short RT_{60}. It is clear that the classic measurement of the time it takes for sound to decay to a level 60 dB down from where it started is not a good predictor of the subjective sense of how reverberant a room feels. This *subjective* quality of the liveness of a room has come to be known as *reverberance*. Studies have shown that the best predictor of reverberance is the Early Decay Time (EDT). Basically the EDT is the time it takes for the sound to decay by 10 dB, and is sometimes referred to as the T_{10}. Reverberance is an important consideration for churches, especially for those churches were there is some sort of music performed, for example churches with pipe organs or choirs.

Actual envelope time curves look like those shown in Figure 21.8. Notice the initial time gap. In this graph the buildup of the reverberant field is obscured by a number of reflections. If we use the same data as shown in Figure 21.8 we can examine a few of the important aspects of the ETC. Examine Figure 21.9: There are

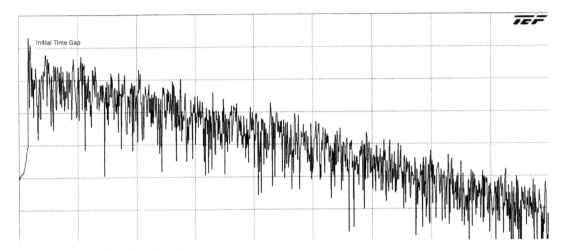

Figure 21.8 Actual Envelope Time Curve.

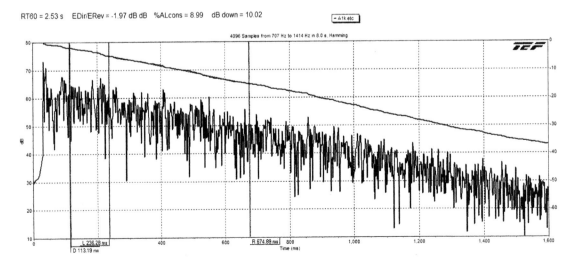

Figure 21.9 EDT and C80.

three vertical lines that are cursors. The smooth line above the ETC is an integration of the curve. The ETD is found by identifying the start of the reverberant field and moving along the integration until the level drops 10 dB. This is the area of the curve bounded by the second and third vertical cursors.

Suggested Listening at www.sound-of-worship.com
Go to the Reverberation Examples section of the website to hear simulations.

Clarity (C80)

Performed music requires a certain amount of clarity if it is to be enjoyed. Researchers have determined that the best predictor of the subjective quality of the clarity of music is the ratio of early to late energy. C80 is the physical dimension associated with the subjective experience of clarity. It is defined as the logarithmic ratio of the energy arriving at a listener in the first 80 ms, to the late sound energy, or the sound arriving after 80 ms.

THE MATH

Clarity or C80 is the logarithmic ratio of the energy in an impulse response before time t_e and the energy after time t_e. For music clarity, C80 t_e is 80 ms. for speech t_e is 50 ms.

$$C_{80} = 10Log \frac{\int_{0}^{80ms} h^2(t)dt}{\int_{80ms}^{\infty} h^2(t)dt} \qquad C_{50} = 10Log \frac{\int_{0}^{50ms} h^2(t)dt}{\int_{50ms}^{\infty} h^2(t)dt}$$

Some have questioned the use of the limit of infinity in the determination of the Clarity factor. Hak and Vertegaal explored this question and determined that a practical limit of the point where the reverb curve intersects with the noise floor can be chosen with negligible error.[3]

Graphically, it looks like Figure 21.9, an Envelope Time Curve (ETC). This measurement was made in a large church. The time scale is from 0 to 1.8 seconds. The first cursor is set by the analyzer to a point 80 ms after the direct sound. C80 is the ratio of the amount of energy in the first 80 ms, to the energy from 80 ms until the decay reaches the noise floor. In this particular location in this church, the C80 read at the top of the graph, is –1.9. C80 numbers that are less than 0, or negative, indicate that at those listening positions there is more reflected than direct energy. Sound experienced in these locations will exhibit less clarity. When the C80 is a positive number, this indicates that the ratio of direct to reverberant energy is positive. A listener who receives more direct than reflected energy is going to experience a sense of clarity that will diminish as the ratio of direct to reflected energy is reduced. C80 is generally thought of as relating to the clarity of music.

Clarity of speech is a closely related parameter but is measured over the first 50 ms rather than the first 80 ms. It is known as C50 and the subjective experience is called distinctness. In concert halls these numbers are directly manipulated by using reflective surfaces that can alter the ratio of early to late energy rather easily. In churches this is often accomplished with the sound system, but it is certainly possible, indeed desirable, to design the acoustic space so that acoustic events can be enjoyed with an appropriate amount of clarity. Generally, in churches positive C80 and C50 ratios are preferred.

Warmth (BR)

Earlier we showed a graph (Figure 21.6) of two hypothetical churches, both with average reverberation times of around 2.5 seconds. The primary difference between the two rooms is the relationship between the reverberation time at the low frequencies and the reverberation time at the midrange frequencies. It has been determined that the subjective character of warmth is best predicted by evaluating the amount of low frequency energy as compared to the midrange. The formula is as follows:

$$BR = \frac{RT_{125\,Hz} + RT_{250\,Hz}}{RT_{500\,Hz} + RT_{1000\,Hz}}$$

Listener Envelopment (LEV)

Listener Envelopment is that quality of a room where the listener feels surrounded and enveloped by the sound. This is generally thought to be a desirable trait of concert halls. At the time of this writing, however, the way to best manipulate LEV is still a topic that is under investigation. Clearly in some churches the property of listener envelopment is also a desirable thing. LEV seems to be most closely tied to the LF_L. LF stands for lateral energy fraction and is the fraction of the energy arriving from lateral sources as compared to the energy arriving from all other sources. This is related to the IACC (InterAural Cross Correlation), which is a measure of the dissimilarity of sounds arriving at each of the ears. The more dissimilar, the more the listener will experience a sense of being enveloped in the sound field. The LF measures a similar parameter as it compares lateral energy, which is beneficial in the creation of envelopment with energy from other locations some of which (especially vertical sources) will be deleterious to the creation of envelopment. LF_L is the lateral fraction measured at octave bands from 125 to 500 Hz.

End Notes

1. Sabine, W. C. (1922). *Collected Papers in Acoustics.* Harvard University Press.
2. Intelligibility, large room frequency, etc.
3. Hak, Constant and Vertegaal, Han, *What Exactly Is Time Infinity for Acoustical Parameters?*, ICSV16, July 2009, Krakow, Poland.

22

NOISE AND ISOLATION

Noise is a problem that every church must deal with regardless of its worship style or form. Noise is basically unwanted sound. It is important to realize that the definition of noise includes a subjective component. Noise is *unwanted* sound. Noise can be any sound that intrudes on someone's space. It can be sound that in another context may be deemed desirable. Noise can originate outside a structure or from within. It can be a by-product of the congregation itself. In any case, in all styles of worship we looked at in the previous section, unwanted sound will always be viewed as a hindrance. Unlike other aspects of acoustics, noise control or abatement is very difficult and expensive to achieve after a building is built. If noise control is built into the structure or design it will not be nearly as costly.

There are two ways that noise, or any sound for that matter, gets into and propagates through a space. Unlike light, sound must have a medium for it to propagate. Sound will travel from its source through the air or through a structure, often both. Both airborne and structure-borne sound are easy to control in theory, but in practice both can be quite challenging, as *every* possible path must be accounted for. With airborne sound, we must

Sound of Worship. DOI: 10.1016/B978-0-240-81339-4.00022-3

achieve virtually airtight conditions if we are to achieve isolation. Structure-borne sound requires mechanical decoupling of the source from the structure.

Consider the following scenario. The First Baptist Church is planning a new building. The design calls for a nursery to be built, as they often are in Evangelical churches, right off the back of the sanctuary. The worshippers in the sanctuary do not want to hear the infants crying—fussing babies would be considered noise to all present with the possible exception of the parents! However, caregivers don't want to miss the entire service, so a window is installed in the wall between the sanctuary and the nursery so the caregivers can observe the service and hear by means of a loudspeaker system installed in the nursery. After the dust settles from the construction, and services are held in the new church, a problem emerges. The worshippers sitting in the back four rows can clearly hear the babies crying as well as the caregivers carrying on conversations in the nursery. The architect is stymied. The walls are appropriately filled with fiberglass insulation and the window was one of those fancy sound-proof ones. How could the noise be getting into the sanctuary? An investigation revealed three paths that sound could take. One was around the glass. If a window is not fully sealed into the frame it will not stop airborne sound. The contractor had not been careful enough with the installation, figuring the trim would cover the gaps around the window. The second path was under the wall plate. The plate was not sealed to the floor and the drywall was cut shy of the floor, leaving an eighth of an inch gap, covered, of course, by the baseboard. Vinyl baseboard does a good job of concealing gaps, but those gaps are all potential air leaks, and therefore sound leaks. The third offending path was an electrical outlet placed in the common wall between the nursery and sanctuary. The electrician had used back-to-back electrical boxes to feed both rooms from one drop, resulting in a significant hole right through the wall. It is very important to note that fixing any *one* of these flanking paths may result in virtually no improvement. Fixing any two might yield a minor improvement. *All* flanking paths must be found and sealed if airborne sound is to be controlled. Other common paths for airborne sound include traveling over a demising wall that is not sealed to the deck, sound leaking through penetrations through walls (or floors or ceilings), for example penetrations to allow for sprinkler systems. HVAC ducts are often the biggest culprits. They can be very effective speaking tubes!

Noise can also intrude into a space through the structure. Sound in the form of vibration will travel very efficiently through solids. The speed of sound in air is approximately 334 meters per second (m/s) (1130 feet per second (ft/s)). The speed of sound in common building materials such as gypsum board is on the order of 3000 m/s. The good news is that for acoustical sources of sound, the transfer of energy from air to structure is very inefficient. A crying baby is not likely to create enough acoustic energy for the sound to travel throughout the structure and cause a problem. However, a simple box fan sitting on a wood floor can often be heard in the apartment below as the vibration is transmitted directly through the feet of the fan to the floor. Once the vibration is in the wood of the floor it will travel to the joists and then on to whatever is attached to the joists and ultimately to the whole structure. This is a much more efficient transfer of energy than is possible through the air. We only have to lift the fan off the floor by an inch or so to prove this. Controlling structure-borne noise can require careful engineering and be a real challenge. We must either isolate the source from the structure, or isolate the "receiving room" from the structure. It is most often better to consider isolating the source rather than the receiver. In the case of a church, it would be silly to acoustically float a sanctuary from the structure of the building to get rid of HVAC noise. Isolating the offending HVAC equipment is much more practical. If we were building a recording studio in a building situated in an urban area where the structure of the building was rife with vibration from all sorts of sources ranging from traffic to HVAC to elevators, the only practical solution may very well be to isolate the entire studio by building the room on isolators.

As with airborne noise, structure-borne noise is conceptually easy to control. Simply isolate the source from the structure. Isolation requires that the vibration energy be dissipated in some sort of elastic medium or that the path the vibration takes from the source to the structure be as inefficient and lossy as possible.

In practice this can prove to be very difficult. Many of the mistakes that we see in the field are not conceptual mistakes. Rather, they are mistakes in execution.

A small Evangelical church in Florida builds a new church. The designer, wishing to save money on HVAC ducting, located the mechanical room directly behind the rear wall of the sanctuary. They took pains to seal the duct penetrations and of course they mounted the HVAC unit on isolators. Why then was the HVAC clearly heard in the sanctuary? It could be argued that the location of the HVAC unit so close to the sanctuary was a bad design decision. The primary reason for the HVAC noise leaking into the sanctuary was because the contractor used the wrong isolators. The isolators were far too soft for the weight of the unit and were bottomed out, resulting in no isolation as they were unable to dissipate any of the energy at all. The same thing can happen if the springs or rubber pads are too stiff. If there is no movement or compression of the isolators there will be no isolation. Most problems seen in the field are much more complex. Figure 22.1 illustrates some of the more common ways that vibration and noise can get from one space to another.

Figure 22.1 represents a common situation with an HVAC system located adjacent to a room intended for some use where the noise from the HVAC unit would be a problem. Each of the numbered arrows represents a path that sound can and does take to get from one room to another. Please note that all the arrows emanating from one duct also apply to the other duct. They are not shown for clarity.

- Arrow 1 represents the vibration that travels through the material of the duct itself and is then radiated into the receiving room. This can be eliminated or dramatically reduced by using canvas or rubber couplings between the HVAC unit and the duct work.
- Arrow 2 represents the noise created by the diffusers themselves. This noise is created by the turbulent air moving through the diffusers. This self-noise can be improved by changing the diffuser design. Most quality HVAC component manufacturers

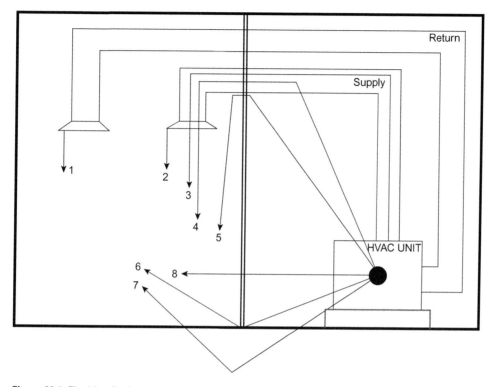

Figure 22.1 Flanking Paths.

will include a noise specification at some face velocity, for example NC 20 (explained in next section) at 400 CFM.

- Arrow 3 represents the sound that originates in the fan and compressor and simply travels through the duct. It is often equally present on both supply and return ducts. This noise can be reduced by the use of duct silencers and duct liner.
- Arrow 4 represents a path that is often overlooked. It is the noise that emanates from the unit and then breaks into the duct, then travels through it just like arrow 3. This break-in noise can also be from other noise sources. It can be controlled by insulating the ducts on the inside as well as the outside.
- Arrow 5 represents the sound that will travel through any cracks left open around the duct penetration of the partition.

This is controlled by making sure all cracks are sealed using a nonhardening sealant.

- Arrow 6 represents the sound that travels under the wall if the floor plate is not sealed to the deck. To eliminate this path, use a nonhardening sealant under the plate and under each layer of drywall.
- Arrow 7 represents the path the vibration will take, traveling through the structure and radiating into the receiving room. Note that the noise will not only be radiated from the floor. The vibration will be radiated from all the surfaces in the room. It can be reduced or eliminated by properly mounting the HVAC unit on isolators appropriate for the weight of the unit.
- Arrow 8 represents the sound that simply travels through the wall. This can be improved by adding insulation into the stud cavity, by adding mass to the wall, or by using materials such as QuietRock® or devices like USG Resilient Channel® to improve the transmission loss of the partition.

Noise, like all forms of sound, is quantified using instruments that measure sound pressure at some frequency or band of frequencies. As mentioned, noise by definition is subjective, so making a correlation between an objective measurement and a subjective outcome can be difficult. Just how quiet is quiet? In addition, even the most powerful instruments available at this writing have trouble distinguishing between sound that is wanted and that which is unwanted. There are, however, some tools and standards that can help in the quest for quieter spaces.

Noise Criteria

The Noise Criteria (NC) and its various permutations, the PNC, NCB, and NR are all ways of quantifying a permitted noise level. The NR curve is the standard in Europe, the NC is the original standard suggested by Beranek[1] in the 1950s. It was replaced in the 1970s by the Preferred Noise Criterion. Figure 22.2 shows a graph of the PNC collection of curves. Each curve represents a criterion. A noise measurement is plotted on the template and the resulting PNC can then be read. In Figure 22.2 the measured

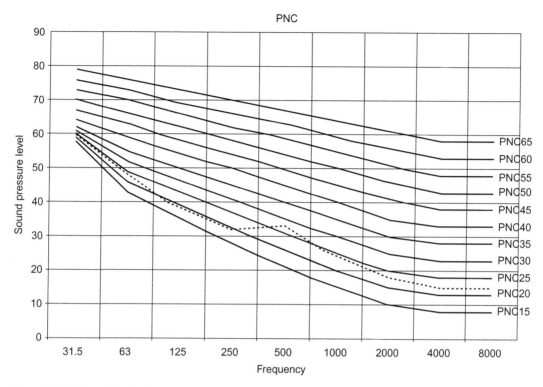

Figure 22.2 Preferred Noise Criterion, PNC 30.

PNC value (shown by the dashed line) is 30 as the measurement exceeds PNC25 but not 30. This is a convenient way of specifying an acceptable level for background noise. It is generally accepted that churches should meet a PNC of 20 to 30, 20 being quieter and more expensive to achieve. PNC 20 is a good target for Sacramental churches.

The PNC is also a very good tool for specifying the permissible levels of systems within a building. HVAC systems are most often the primary noise culprits in churches. Specifying that the HVAC may not produce levels that exceed a PNC 25 measured in the sanctuary is a way of holding the mechanical engineer to a verifiable objective standard.

Transmission Loss

There are two ways to build walls with high transmission loss. We can use a brute force approach and use mass to control the transmission of sound. In theory if the mass of a panel is doubled, there should be a 6 dB improvement in the transmission loss. In reality, due to a number of causes the increase is more like 4.4 dB for every doubling of mass. Likewise for every doubling of frequency above a certain point there will be a 6 dB improvement in transmission loss.

THE MATH: MASS LAW

The math for the Mass Law is pretty straightforward. The formula $R = 20\log(fm) - 47$, in dB, where R is the reduction in sound level through a partition; f is the frequency of the impinging sound in Hertz; and m is the mass per unit area in meters^2. This is a simplified formula based on the assumption that the characteristic impedance of the air is around 410 rayls for 20°C at one atmosphere. Strictly speaking the mass law only applies to nonrigid or limp partitions. Most real world walls involve a complex interplay between mass, stiffness, and dampening. The mass law continues to predict the behavior until the critical frequency is reached. The critical frequency is where the wavelength of the bending waves in the structure match the frequency of the impinging wave. Unlike sound in the air, bending waves in material travel at speeds determined by the frequency of the wave. For the frequencies above a certain critical frequency, there will be an angle of incidence for the impinging wave where the bending wave and the impinging wave will be at the same wavelength. When this happens it is known as coincidence. At coincidence, the transmission loss can be greatly reduced even at voice frequencies, especially in materials like glass. The coincident frequency can be found with the following equation:

$$F_c = c^2/(1.8 * h * vL * \sin^2(a))$$

where c is the speed of sound in m/sec, h is the panel thickness in meters, vL is the longitudinal velocity of the sound in the partition, and a is the angle of incidence of the impinging sound.

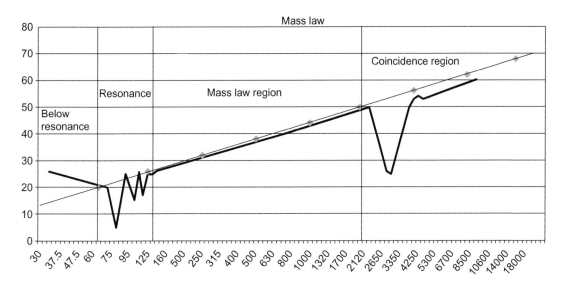

Figure 22.3 Mass Law Chart.

The chart in Figure 22.3 illustrates the performance of a typical mass-only partition. An example of a mass-only wall is a brick or CMU wall. Every material has a natural frequency or frequency at which it will vibrate once set into motion. If we call this natural frequency Wn, and the excitation frequency W, when $W \ll Wn$, generally very low frequencies, hopefully below the audible range, the stiffness controls the behavior of the partition. In this region, the first region in Figure 22.4, there is a 6 dB drop in loss per doubling of frequency. When $W = Wn$, the system is at resonance and it is the damping that controls the system. At these frequencies there is very little if any transmission. When $W \gg Wn$, the mass of the system dominates. In the mass-controlled region, the transmission loss will increase by 6 dB for every doubling of frequency. At some frequency known as the Critical Frequency, there is a significant dip in the TL. This coincidence dip is a function of the make-up of the material, its surface density, stiffness, and thickness and the angle of incidence of the impinging wave.

A better way to achieve higher transmission loss is to use a complex partition. These partitions are really mass–spring–mass

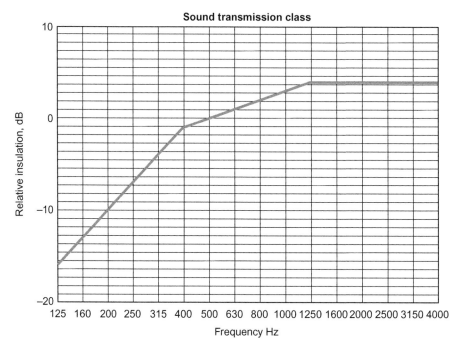

Figure 22.4 Sound Transmission Class.

combinations. Even a simple stud wall with a single stud and a layer of drywall attached to each side is a sort of complex wall. The two layers of drywall are two masses and the air between them is effectively a spring. Of course there must be studs in the wall, and the studs reduce the transmission loss considerably. Because they have some resilience, metal studs offer higher transmission loss values than wood studs with the same number and arrangement of layers of drywall. Adding a resilient channel such as the USG RC improves the transmission loss by improving the spring in the mass–spring–mass system. Staggered stud and double stud walls perform even better, of course taking up more space in the process. Adding fiberglass to the cavity does two things; it dampens the resonance, improving the low frequency transmission loss, and it adds some mid and high frequency absorption to the cavity, improving the TL in those frequency ranges as well.

Sound Transmission Class

How well a partition blocks or prevents the transmission of sound through it is rated using the Sound Transmission Class. The STC is a standard defined by ASTM E 413 *Classification For Rating Sound Insulation*. For a full description of the standard and the measurement methodology, refer to the standard. It is most important to remember that the STC is a laboratory rating of a partition or device. ASTM E 413 also makes provision for determining the FSTC or Field Sound Transmission Class. In an actual application the device or partition will likely not perform as well as it did in the lab. An acoustical door, rated at a STC 50 in a standard laboratory test, can be expected to perform at approximately FSTC 45 installed. Figure 22.4 is the reference contour graph that is used to determine the Sound Transmission Class. The reference contour is used in following manner.[2]

Measure the device (partition, window, door, structure, etc.) according to ASTM E90. Plot the data.

Fit the reference contour to the data such that some of the measured data are less than the reference contour. At each frequency, calculate the difference between the reference contour and the measured data. Deficiencies are defined as those data points that fall below (have a value less than) the reference contour. Only the deficiencies are used in the fitting procedure. Continue to adjust the reference contour until the most stringent of the two conditions are met:

1. The sum of the deficiencies is equal to or less than 32 dB.
2. The maximum deficiency at any one frequency does not exceed 8 dB.

To demonstrate, consider a window assembly that is tested according to ASTM E90. The data are shown in Figure 22.5.

By either using a transparency or by calculation, follow the steps just listed. The graph will look like Figure 22.6.

Going back to the example of the Baptist church building a nursery right off the sanctuary, assume that the ambient noise in the sanctuary is measured at PNC 30. If a crying baby is measured

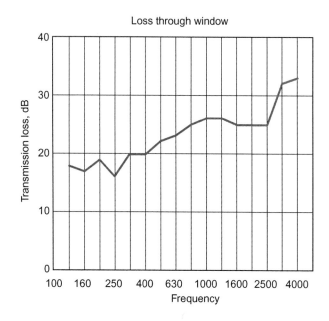

Figure 22.5 Transmission Loss of Window Assembly.

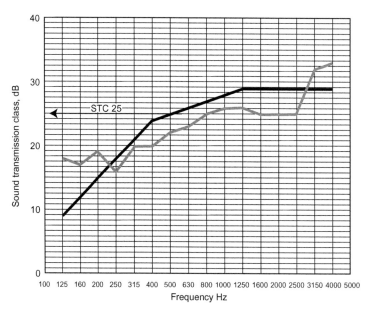

Figure 22.6 STC of Window Assembly—STC 25.

at 90 dB SPL at 500 Hz in the nursery, in order for the crying to be inaudible, the walls and window structure would have to meet a STC of around 60 if the crying is to be at the same magnitude as the ambient noise of NC 30. In reality, the crying would have to be *below* the ambient noise in order to be inaudible, meaning that the STC would have to be greater than STC 60. Of course in an actual service there are rarely times of complete quiet so the sounds of the service would help mask the crying to some extent. A STC 60 is a fairly stringent specification. It is difficult to build systems that meet STCs above 50 using conventional commercial construction.

Recently we were working on a church build-out where one of the desired elements was three practice rooms to be used by various youth-oriented bands. Some of these bands were documented rehearsing at levels that exceeded 110 dB SPL! The architect of course wanted to provide a sound-proof space for these bands and locate it in the middle of the church office space. There are a number of problems with this plan! In addition to the very real issue of hearing damage, the term "sound-proof" is rarely if ever used by reputable acoustical consultants. We can't sound-proof a room. We can improve the transmission loss, but true sound-proofing, if possible, is enormously expensive. Looking at the numbers, if the band is actually playing at around 110 dB, the rooms including the doors and HVAC systems would have to contain at least 75 dB of that to bring the noise down to a level commensurate with the office background noise of around NC35. Even with that kind of attenuation the band would still be audible! Construction of entire systems—not just walls but floor systems, ceiling systems, doors and HVAC systems—that would provide for the isolation necessary would be prohibitively expensive.

Impact Isolation Class

In multistory buildings it is often desirable to design floor systems such that occupants of one floor cannot hear people walking around on the floor above. It is not enough to simply state in a specification that hearing footsteps is not acceptable! ASTME989 sets forth a standard known as the Impact Isolation Class (IIC) and also proscribes a measurement protocol. The measurement of IIC requires the use of a tapping machine.

Figure 22.7 Tapping Machine.

A tapping machine (Figure 22.7) consists of a motor-driven cam shaft and five weights of a specific mass. The cam shaft lifts the weights by a specified amount and then permits them to free fall to the surface under test. In this way a standard, repeatable tapping stimulus is introduced on one floor (the sending room) and can be measured in the floor below (the receiving room). Because of the many flanking paths that are often present in commercial construction, field measurements of IIC are generally 5 to 6 dB worse than the corresponding laboratory measurements. Like the STC, the IIC is a one number rating that is found by fitting measured data under a reference contour. Because the IIC is a measure of structure borne-vibration, the calculation of the IIC requires that the tapping machine is placed in five locations and the results are averaged. In this way the contributions from the structural elements like floor joists are averaged out. The acoustics of the receiving room are also taken into account. The most effective way of improving IIC after construction is complete is the addition of carpet and padding on the floor. If carpet is not an option, like in the case of a kitchen or bathroom, it is very hard to improve IIC without major construction or redoing the floor.

The IIC standard was first written in 1977. It is certainly better than having no standard at all, but it is in dire need of update. One of the severe limitations to the IIC spec is that it only deals with frequencies down to the octave of 100 Hz. Furthermore, the tapping machine does not accurately predict what actual transmission will be like. There is a big difference between the sound of 1.1 lb hammers falling 1.5 inches and the sound of a 175 lb person walking in hard-soled shoes.

End Notes

1. Preferred Noise Criterion curves, from Beranek, LL., Blazier, WE., & Figwer JJ., (1971). Preferred noise criterion (PNC) curves and their application to rooms. *Journal of the Acoustical Society of America*, 50, 1226, used by permission of the authors.
2. ASTM E 413-87, 1999, p. 2.

23

INTELLIGIBILITY

Simply put, intelligibility is the ability to understand the spoken word. It is not, strictly speaking, a function of the acoustics of the room, but acoustics do play an important role in intelligibility. Intelligibility is not to be confused with audibility. Audibility is a measure of hearing whereas intelligibility is a measure of *understanding*. Everyone involved in the design of churches or of sound systems for churches needs to understand how to optimize intelligibility, even though some worship styles will not value intelligibility as much as others do.

For some churches, most notably those that worship in the evangelical style, intelligibility is the most important design criterion of the whole project. As we have seen, for the Evangelical church, preaching the Word is everything. Many see the scripture passage in the book of Romans as the definitive statement about church: "Romans 10:14 How, then, can they call on the one

Sound of Worship. DOI: 10.1016/B978-0-240-81339-4.00023-5

they have not believed in? And how can they believe in the one of whom they have not heard? And how can they hear without someone preaching to them?"[1] The ability to understand the spoken word is particularly important to these "preaching" churches as there is no liturgy or ritual that remains the same from week to week, to help fill in the missing parts. Back in Chapter 5, there is a story of an elderly Irish man who approached a consultant while he was working on a sound system in a Catholic church. He said, "Ach laddie, I dunna need to hear the prayers. I've been hearing the prayers since I was twelve. But I suppose it would be nice once to hear the homily!" What he was saying is that he had long since memorized the Mass. As soon as the priest starts with the familiar "The Lord be with you" and the reply of the congregation, "and also with you," the devout Catholic knows what is coming, but in the Evangelical church, what is spoken is to a great extent new every week. Since the Evangelical church sees converting non-Christians as its primary purpose, it expects that there is a possibility of someone attending a church service who has never heard the Gospel. This person not only has no ritual to help fill in the parts, but he or she also has no linguistic context as he or she is likely unfamiliar with the language or the jargon of the Church. It is critical therefore that the technology, whether the sound system or the architecture, not interfere with the message.

How we understand speech is obviously a very complex topic, far beyond the scope of this book. However, in the last 30 years or so there has been significant research into predicting the performance of a sound system in a room with respect to intelligibility. We are now able to predict whether or not a typical listener will be able to understand the spoken word, and after the room is built and the system installed, we are able to document with a degree of certainty the ability to understand the spoken word.

As soon as the telephone became a commercial success, researchers began looking into the question of intelligible speech and what factors would inhibit the ability to understand speech. Evaluating intelligibility broadly divides into two approaches: the statistical approach that uses subjects as listeners, and the objective- or measurement-based methods.

Subjective/Statistical Methods

Lochner and Berger proposed a standardized word list for statistical evaluation of intelligibility of reinforced speech in 1959.[2] Their approach consisted of a reader who is a natural speaker of the language used, reading lists of words or sentences. Listeners then would write down what they heard and the results would be compared to the original and an intelligibility score would be determined. As work continued in this field it became clear that this is indeed a complex task. Over the years different approaches were developed to try to directly measure the ability to understand speech.

The Diagnostic Rhyme Test (DRT) uses 96 pairs of words that differ only in the initial consonant sound; for example, *reed–deed* or *veal–feel*. The listener is given a score sheet with the pairs written out. One word of each pair is spoken by a reader. The listener marks which one he or she thinks is correct. The Modified Rhyme Test uses a similar approach to the DRT but has a matrix of 300 words testing both the initial and ending consonant sounds.

Around the time of the second World War, Harvard University developed a set of phonetically balanced monosyllabic words now known as the PB word list. There have been tests developed using a wide variety of techniques including lists made up of nonsense syllables and using entire nonsensical sentences. All the direct or subject-based techniques require some amount of training to use correctly and require a fair amount of statistical processing to extract meaningful information about a specific sound system/room transmission chain.

Measurement Methods

Articulation Index

The Articulation Index (AI) is a very accurate method of evaluating intelligibility, initially developed for evaluating telephone lines. It uses measurements of the spectrum of the signal as well as the spectrum of the noise. It is not well suited for use in sound systems in rooms. In 1997 the standard was revised and the Speech Intelligibility Index was described. The SII is a reworking of the AI and is quite accurate but cumbersome. Currently few if any commercially available analyzers measure SII.

%ALCONS

The pioneer of intelligibility prediction research is V.M.A. Peutz, a Dutch researcher who published the first significant paper on the topic in 1971. The work, entitled *"Articulation Loss of Consonants as a Criterion of Speech Transmission in a Room,"*[3] was the first to suggest a criterion for objectively quantifying intelligibility before a system or room was built. Engineers had been using word lists to evaluate intelligibility before 1971, but by the time a wordlist could be read in a space it was often too late to do anything about a poor intelligibility score. Furthermore the elements that contribute to poor intelligibility were not well understood. Peutz suggested that for languages that were broadly related to English (German, Dutch, etc.), the ability to differentiate between words and therefore the ability to understand a word is tied to the ability to clearly hear the consonant sounds. Peutz coined the acronym %ALCONS (Percent Articulation Loss of Consonants), and suggested that any more than 15 percent of consonant loss and the speech will be unintelligible. Most sound system designers however use a %ALCONS rating of 7 to 10 percent as the cutoff, citing 15 percent as unacceptable.

THE MATH: %ALCONS

The %ALCONS equation (note in metric units the constant 656 becomes 200):

$$\%AL_{cons} = \frac{656(D_2)^2(RT_{60})(n+1)}{VQ}$$

where

D_2 = distance between the loudspeaker and furthest listener in feet

RT_{60} = the reverberation time in seconds

V = volume of the room in ft^3

Q = the directivity of the loudspeaker (dimensionless)

n = the number of loudspeakers directly contributing to the sound at the listener

We will not show the derivation of this equation here; refer to Peutz's paper.

It is instructive to look at the variables of Peutz's equation to understand the role that each plays in intelligibility. This equation suggests that if you know the distance between the loudspeaker and the listener D_2, the reverberation time, RT_{60} (see Chapter 21), the number of loudspeakers that contribute to the listener (n), the directivity of the loudspeaker (Q), and the volume of the room (V), you can predict what percentage of the consonant sounds will be lost. Clearly, as (Q) or loudspeaker directivity goes up (i.e., as the speaker becomes more directional), the articulation scores get better. As the reverb time increases in a given space the articulation scores get worse. %ALCONS is a straightforward tool useful in predicting intelligibility, especially in simple systems, but it is tricky to measure.

One of the difficulties in using %ALCONS is that as systems become complex, it is almost impossible to know the actual Q of a loudspeaker cluster, to say nothing of a distributed system. Still, Peutz's work shows that that the ability to understand speech is a special case of a signal-to-noise problem, where signal is the speech, and noise is all unwanted sound. Elements that increase signal—making D_2 smaller or making Q larger—result in improved intelligibility. Increasing the reverberant level or reducing the directivity results in degraded intelligibility. One drawback to the Peutz equation is that it does not account for noise that is uncorrelated to the signal. Reverberation and discrete reflections are essentially the direct sound repeated and reflected so in that sense reverberation is correlated to the signal. The noise from the air conditioner, for example, has nothing to do with the signal and as it turns out affects intelligibility in a different way. Still, the Peutz equation and %ALCONS remains a useful tool for designing and specifying systems.

STI

As Peutz was developing %ALCONS as a method of *predicting* intelligibility, two other Dutch researchers, Steeneken and Houtgast, were working on a method of *measuring* intelligibility directly. Their technique, called the Speech Transmission Index (STI), is mathematically much more intensive and since it is now possible to create virtual rooms and model the response of loudspeaker systems with sufficient accuracy, STI measurements can be made on *virtual* systems in rooms before they are built. In many parts of Europe STI has become a standard for specifying the performance of a sound system in a room. It is based on the fact that there are natural amplitude modulations that occur in speech. The Modulation Transfer Function (MTF) has been around for a long time. The MTF is used to measure how well the amplitude modulation of a signal, that is the variation of intensity with time, is preserved through a transmission chain, in this case a room. M.R. Schroeder showed that the MTF could be derived from the impulse response of a system.[4] The theory suggests that if the modulations are interfered with, for example if the trough of the modulation is "filled up" with noise or reverberation, there will be a degradation of intelligibility. Graphically, MTF can be depicted in the following way. Figure 23.1 shows a voice waveform. The peaks and dips in amplitude are the natural modulations in the voice.

Figure 23.2 shoes a waveform of noise. Notice that it has almost no modulation.

Now imagine that you superimposed the noise on the speech, as in Figure 23.3. The modulations in the speech are significantly reduced, resulting in a severe loss of intelligibility.

Figure 23.1 Voice Waveform.

Figure 23.2 Noise.

Figure 23.3 Noise Superimposed on Speech Waveform.

THE MATH: SPEECH TRANSMISSION INDEX

Schroeder showed that the MTF can be calculated from the impulse response of a system. The formula

$$m(\omega) = \frac{\int\limits_0^\infty h^2(t)e^{-i\omega t}\,dt}{\int\limits_0^\infty h^2(t)\,dt}$$

is the Fourier transform of the square of the impulse response, divided by the total energy of the impulse response.[5] The output is divided by the input to find the MTF at a given frequency. Once the MTF is known, the STI can be calculated by first finding the Transmission Index (TI). $TI_i = \frac{x_i + 15}{30}$

where X is a number representing the signal-to-noise for the modulation signal and is limited to ±15. The Octave Transmission Index is then found for each octave band:

$$OTI_n = \frac{1}{14}\sum_{i=1}^{14} TI_i \quad n = 1,2,\dots 7\,.$$

Suggested Listening at www.sound-of-worship.com
 Visit the Intelligibility section of the site.

Finally, the STI rating is found using the following equation:

$$STI = \sum_{n=1}^{7} \alpha_n OTI_n - \sum_{n=1}^{6} \beta_n \sqrt{OTI_n \times OTI_{n+1}}$$ where α and β are

weighting factors experimental determined for male and female speech.[6]

STI measurements are made in seven octave bands from 125 Hz to 8 kHz, looking at typical speech-like modulations in each. Each octave is given a score ranging from 0 to 1 (0 is no intelligibility, 1 is no degradation) and then the seven results are compiled into a single STI rating, with some of the octave bands given preferential weighting. The STI rating corresponds well with results from subject-based word lists. STI ratings from 0.6 to 1 are considered good to excellent, 0.5 is considered fair, and below 0.5 is poor or low intelligibility (see Figure 23.5). When STI was first suggested, there was not sufficient computer power available to make the measurements practical. An alternative method called RaSTI (Rapid Speech Transmission Index) was developed as a faster subset of STI. RaSTI only looks at the speech modulations in the 500 Hz and 2 kHz bands and is indeed much faster to use, but it is nowhere near as accurate. One of the leading intelligibility experts, Peter Mapp, described the response of an admittedly contrived sound system (figure 23.4) that delivers virtually no intelligibility and yet when measured with RaSTI it yields an almost perfect score.[7]

Now with the advent of faster computers, the full STI measurement can be performed in a few minutes, obviating the need for RaSTI. It is important to remember that STI does not measure intelligibility directly, but it does measure aspects of the transmission vehicle, if you will, that do correlate very well with intelligibility.

STIPA

STIPA (Speech Transmission Index for PA) was developed in 2001 to try to overcome some

Figure 23.4 System Response Yielding Excellent RaSTI Values, But Poor Intelligibility. (After Mapp).

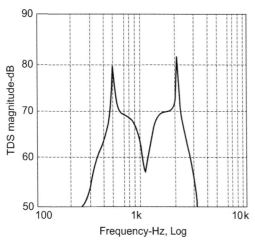

of the problems associated with STI and RASTI. It uses the MTF as does STI, but instead of relying on measuring the impulse response STIPA uses a proprietary recording of a stimulus that mimics the modulation characteristics of speech. This allows for the development of some very simple user friendly intelligibility meters. The user simply plays a sound file from a CD or other recording media. While the sound file is playing the user points a handheld device at the speaker/room under test. Usually the handheld device then indicates a subjective rating of the system from excellent to poor.

These two concepts, the articulation loss of consonants and the speech transmission index, have furthered our understanding of how we understand speech through an amplified system in a room. In no particular order, the most important or primary aspects of a sound system/room that impact our ability to understand speech are:

Room Reverberation Time: In general, the higher the RT_{60}, the lower the intelligibility.

Ratio of Signal (desired sound)-**to-Noise** (undesired sound): The higher this ratio the better.

System bandwidth and frequency response: Nonlinear systems and systems with limited bandwidth will often have poor intelligibility.

Distance between the listener and the primary source (loudspeaker): The further away from a source, the more likely that the signal-to-noise ratio will decrease.

Directivity Index or "Q" of the loudspeaker (another signal to noise modifier): In reverberant spaces, low Q speakers will deliver lower intelligibility than high Q loudspeakers if the high Q loudspeakers are pointed at the audience.

Number of loudspeakers or loudspeaker clusters in the space: Fewer is better as this too impacts signal-to-noise for any given listener. Even the most directional speakers systems

Subjective	STI	% Alcons
BAD	0.2	57.7
	0.24	46.5
	0.28	37.4
	0.32	30.1
	0.36	24.2
Poor	0.4	19.5
	0.44	15.7
	0.48	12.7
	0.52	10.2
Fair	0.56	8.2
	0.6	6.6
	0.64	5.3
Good	0.68	4.3
	0.72	3.4
	0.76	2.8
	0.8	2.2
	0.84	1.8
	0.88	1.4
Excellent	0.92	1.2
	0.96	0.9
	1	0

Figure 23.5 STI and %ALCONS Values and Subjective Scale.

contribute to the "noise" for those listeners not in the direct coverage.

Ratio of direct to reflected sound: There is still some debate as to when reflected sound becomes deleterious; most use either 35 ms after the direct sound or 50 ms after the direct sound as the onset of reflected energy that negatively impacts speech. This is different than noise because although in the broad definition of noise, reflected energy is indeed unwanted sound, reflected energy is mostly correlated with the direct sound; that is to say, the reflected sound field is made up of delayed images of the direct sound. Noise usually is unrelated to the direct sound.

There are other factors that impact the intelligibility in churches (these might be considered secondary):

System Distortion: Lower is better—interesting to note that some forms of distortion actually increase intelligibility where as other forms of distortion can degrade intelligibility. Although some forms of distortion might be shown to improve intelligibility, distortion will rarely be viewed as desirable overall.

System Equalization (see bandwidth and frequency response, earlier): Improper equalization can exacerbate an already poor response.

Presence of Early Reflections: Reflections that occur within a few ms of the direct sound will result in severe frequency response degradations which in turn can reduce intelligibility.

Presence of late, isolated (often focused) reflections: Reflections that arrive at a listener late (i.e., after 70 ms) especially if they are high in level relative to the direct sound often are heard as slaps or echoes and dramatically reduce intelligibility. These are often the result of architecture that focuses reflections to some section of the room.

Direction of direct sound arriving at listener: Anything that makes the listener work harder to understand will affect intelligibility. In a church if the loudspeaker is on the side wall, but the talker is standing at the pulpit, there will be a subliminal confusion, and this can have a negative impact on intelligibility.

Human factors: We must also include as important aspects affecting intelligibility the human factors that are not of course part of the design process.

Talker Skill; that is, rate of delivery, enunciation, projection, etc.

Talker microphone technique

Listener skill

Vocabulary shared between talker and listener

We have mentioned before the book the intriguing title *If Bad Sound Were Fatal, Audio Would Be the Leading Cause of Death.*[8] Although it is slightly overstated, the Davises recount in the form of an informal history of Syn-Aud-Con numerous stories of really bad sound. From the point of view of the Church, nowhere are the stakes higher. All too often churches, especially the smaller ones, are victims of bad sound either by (poor) design on the part of the sound people or by architecture that is visually interesting but acoustically devastating, or by the merging of the two. Setting an intelligibility specification is a way of holding all to a standard of performance that will help ensure that the technology will not get in the way of the message.

Standards

As of this writing there are essentially two standards that prescribe ways to quantify intelligibility. The ANSI (American National Standards Institute) has created the Speech Intelligibility Index (SII) and sets forth the standard in ANSI S3.5-1997. An earlier standard, ANSI S 32. 1989, covers the use of phonetically balanced (PB) word lists. The other is from the International Electrotechnical Commission, the IEC 60268-16. This standard, used throughout Europe, includes STI, RASTI, and STIPA.

In 2010 there is a new development in the United States. The National Fire Protection Association (NFPA) has just adopted a new standard, the NFPA 72-2010. Mass Notification Systems using voice warning messages instead of, or in addition to, alarms and flashing lights in public buildings have been mandated for a while. The new code stipulates that the voice message must pass intelligibility tests. The code allows for either a subjective test or an objective test to be performed. The subjective test is based on the ANSI S 3.2 -1989 BP word list method. The objective test calls for a STIPA test to be used. A full description of the code can be found in the NFPA 72-2010 Appendix.

End Notes

1. Romans 10:14, New International Version
2. Lochner, J. P. A., & Berger, J. F. (1959). The intelligibility of reinforced speech. *Acoustica, 9.*
3. JAES, v. 19 no. 11, Nov. 1971, pp. 915–919.
4. Schroeder, M. R. (1981). Modulation transfer functions: Definition and measurement. *Acoustica, 49.*
5. Keele, Jr., D.B. *Evaluation of room speech transmission index and modulation transfer function by use of time delay spectrometry.* AES 6th International Conference preprint.
6. Mateljan, I. (2004–2009). *User Manual for ARTA.* version 1.5.
7. Ballou, G. (Ed.), *Handbook for Sound Engineers* (4th ed., p.1239). Focal Press.
8. Carolyn, D., & Don, D. (2004). *If Bad Sound Were Fatal, Audio Would Be the Leading Cause of Death,* 1st Books Publishing, Bloomington, IN.

GLOSSARY

Altar The altar is the table in the chancel that the clergy use for Communion. During the Protestant Reformation, some people felt that the traditional term was theologically misleading. As a result, many people preferred to call it a Communion Table. Anglicans decided that both terms were correct, because it is the altar from which we receive the sacrifice of Jesus Christ, and because it is, literally, the table on which we celebrate Communion. Today, Anglicans and Lutherans generally call it the altar, whereas churches in the Reform tradition tend to call it a Communion table.

Apse If the wall behind the altar (the east wall) is curved, it forms a semicircular area that is called an apse. In ancient times, large church buildings were modeled after a type of Roman public building that had such a wall.

Ambo If there is one speaker's stand in the center of the front of the church, as is typical in churches with a lecture-hall floor plan, it serves the functions of both lectern and pulpit. The word *ambo* comes from a Greek word meaning "both." In common usage, however, ambos are incorrectly called pulpits.

Ambry (or Aumbry) An ambry (or aumbry) is a niche in the wall in a large church. It is generally used for storing various articles that are used in worship.

Baptistery In a Roman house, the household's water source was in the atrium just inside the front door. When early Christians converted a house to a church, that water source became the place where baptisms could take place if it wasn't possible to baptize outdoors. Even though the position of the baptistery was determined by the existing architecture of the house, it took on a symbolic meaning, because baptism is the entrance to the Christian life. Today the position of the baptistery varies. It can be in one of three places: just inside the doors, in the nave in front of the congregation, or behind the chancel.

In churches that usually administer baptism by pouring, the baptistery consists of a stand with a water basin on top. It could be a permanent structure in the front of the congregation or just inside the church doors, or it could be a portable structure that appears only when there is a baptism.

In Protestant churches that administer baptism by immersion, the baptistery is a large tank that is located in the front of the church, either behind the chancel or to one side.

The Catholic Rite of Christian Initiation for Adults calls for baptism by immersion. In newer Catholic churches that are built with this rite in mind, the baptistery is generally an artificial pool with a water pump so there is a continuous flow of water. It can be located just inside the entrance of the church, or in the nave in front of the congregation.

Cathedra The chair on which the bishop sits. It is located in the chancel, often centered behind the high altar. When a bishop (such as the pope) speaks *ex cathedra*, it means he is speaking in his official capacity.

Cathedral Refers to the function of a church, not its architectural style. A cathedral is a church that serves as a bishop's headquarters, so to speak. It's called a cathedral because it contains his cathedra (chair). The city in which the cathedral is located is the bishop's see. In this usage, the word see comes from a Latin word meaning seat. The city is the bishop's see in the sense that a city might be the seat of government.

Chancel   In churches with a historic floor plan, the chancel is the front part of the church from which the service is conducted, as distinct from the nave, where the congregation sits. The chancel is usually an elevated platform, usually three steps up from the nave. In churches with a lecture-hall floor plan, the term sanctuary is often used to mean both chancel and nave because the two are not architecturally distinct. In the historic floor plan, the words chancel and sanctuary are often synonyms.

Chancel Screen   See *rood screen*.

Chapel   A chapel can either be an alcove with an altar in a large church, or a separate building that is smaller than a full-sized church. Chapels have the same function as church buildings and are equipped the same way, but usually they are dedicated to special use. For example, a large estate might have a chapel in which worship services are held for family members, staff, and guests. If a church builds a new and larger sanctuary, but keeps the old one, the old one is often called a chapel.

Communion Table   See *altar*.

East Wall   The wall behind the altar, as viewed from the nave, is the "east wall," no matter what direction you are actually facing. In the past, all church buildings faced east, and it is still the case for eastern Orthodox churches today. A person who enters the church goes from west to east, which symbolizes going from the evil of the present world to the glory of the New Jerusalem to come.

High Altar   A large church may have several altars. The term high altar refers to the main altar in the chancel. Other altars may be located on the sides of the nave or in separate chapels in the same building.

Historic Floor Plan   As viewed by a worshipper seated among the congregation, there are two speaker's stands on either side of the front of the church. The one on the left is called the pulpit, and it is used by clergy to read the gospel lesson and to preach the sermon. Accordingly, the left side of the church is called the *gospel* side. The one on the right is called the lectern. It generally holds a large Bible and is used by lay readers for the Old Testament and epistle lessons. Accordingly, the right side of the church is called the *epistle* side. The communion table stands centered behind the lecterns. If there is enough room, the communion table is placed away from the wall so that the celebrant may face the congregation during communion. The choir may be located behind the congregation, to one or both sides of the sanctuary, or even on the opposite side of the communion table from the congregation. The choir is most often not in direct sight of the congregation. The wall that the congregation faces during worship is called the "east wall," regardless of the actual compass direction, because of the ancient practice, inherited from Judaism, of facing Jerusalem during prayers.

The simplest and easiest shape for a room is a square or rectangle, because it is easier and less expensive to build a straight wall than a curved wall. In the historic floor plan, the chancel is on the short wall of the rectangle. That results in a long aisle and pews in the back that are quite some distance from the front. There are two modern variants on the historic plan; one is to put the chancel on the long side of the rectangle and the other is to make a square room and put the chancel in one of the corners. In these variants, the pews are either curved or placed at angles so that everyone in the congregation faces the chancel. The result is that everyone is closer to the chancel. Orthodox churches also follow this plan, except that they actually do face east, the nave is square rather than rectangular, and there are normally no pews (the congregation stands).

See also *iconostasis*.

Icon   An icon is a highly stylized religious painting on wood. The icon follows detailed artistic conventions, which include the lack of perspective and unearthly colors. The icons are deliberately unrealistic so that they edify faith without causing idolatry. In an Orthodox church, no matter where you look, there's an icon—and that is the whole idea. It is nearly impossible to be in an Orthodox church without thinking spiritual thoughts all the time. The subject and placement of the icons is significant. An illiterate person could learn the whole gospel just by looking around.

Iconostasis   In Orthodox churches, the chancel and the nave are separated by a partition that generally does not reach all the way to the ceiling. It is covered with icons whose subject and placement is significant. It is called an iconostasis—it is essentially an icon stand. The iconostasis has three doors, two on each side, so the clergy can enter and leave the chancel; and one in the middle that, when open, gives the congregation a view of the celebrant and the altar. In Orthodox worship, the nave represents earth, the chancel represents heaven, and the iconostasis is the barrier that prevents us from seeing heaven from earth. The celebrant opens the middle door at appropriate times when heaven is revealed to people on earth.

   The western equivalent is called a *rood screen*.

Kneeler   In churches where it is customary to kneel for prayer, there is often a long, narrow padded bar at the base of the pew in front of you, which can be tilted down for kneeling and tilted up to make it easier to get in and out of the pew. Most often the kneelers are the length of the pew and are used by several people. If you are visiting a church that has kneelers, and you are not accustomed to using them, keep the kneeler in the down position during the service except while someone is passing through. Otherwise someone might attempt to kneel when the kneeler isn't in place. See also *prayer desk*.

Lectern   In churches with a historic floor plan, there are two speaker's stands in the front of the church. The one on the right (as viewed by the congregation) is called the lectern. The word lectern comes from the Latin word meaning "to read," because the lectern primarily functions as a reading stand. In Celebratory churches, it is used by lay people to read the scripture lessons (except for the gospel lesson), to lead the congregation in prayer, and to make announcements. Because the epistle lesson is usually read from the lectern, the lectern side of the church is called the epistle side. See also *ambo* and *pulpit*.

   In some churches, the positions of the pulpit and the lectern are reversed (that is, the pulpit is on the right and the lectern is on the left) for architectural or aesthetic reasons.

Lecture-Hall Floor Plan   As viewed by a worshipper in the congregation, there is one speaker's stand, centered in the front of the church. It is technically an ambo, but is often called the pulpit. It is used by all individuals who are involved in the conduct of the worship service. The choir is seated behind the pulpit, facing the congregation and in full view. There is usually a long kneeling rail between the congregation and the pulpit. If there is a communion table, it is located between the kneeling rail and the pulpit. To receive communion, the congregation comes up and receives the elements at the rail. In some churches communion is served to the congregation in the pews. The kneeling rail often is used for individual counseling and prayer as a response to the sermon or the worship service.

Narthex   The historic term for what might otherwise be called the foyer or entryway of the church.

Nave   In Celebratory churches, the architectural term for the place where the congregation gathers for worship, as opposed to the front part of the church from which the service is led. In Protestant churches, the term *sanctuary* often is used to mean both chancel and nave because the two are not architecturally distinct.

Oratory   A room or a portion of a room that is set aside for an individual to conduct personal devotions. The word oratory comes from a Latin word that means *a place to pray*.

Pew   Originally, Christians stood for worship, and that is still the case in many eastern churches. The pew, a long, backed bench upon which congregants sit, was an innovation of western medieval Christianity. Pews were inherited by Protestants from the Roman Catholic church, and because of their practicality, have spread to some Orthodox churches located in the west.

Prayer Desk   Also called a *prie-dieu*, a prayer desk is a kneeler with a small shelf for books, as in the illustration on the right. In churches where it is customary to kneel for prayer, there might be two prayer desks in the chancel, one for the clergy and the other for the lay leader. Prayer desks are also found in private homes and small chapels.

Prie-Dieu   See *prayer desk*.

Pulpit   In Celebratory churches with a historic floor plan, there are two speaker's stands in the front of the church. The one on the left (as viewed by the congregation) is called the *pulpit*. It is used by clergy to read the gospel and preach the sermon. Since the gospel lesson is usually read from the pulpit, the pulpit side of the church is called the *gospel* side. See also *ambo* and *lectern*.

In some churches, the positions of the pulpit and the lectern are reversed (i.e., the pulpit is on the right and the lectern is on the left) for architectural or aesthetic reasons. In Protestant churches, the pulpit is often the center of visual focus. It also is used as a metaphor for the source of authority.

Rood Screen   A rood screen (also known as a chancel screen) is a partition that separates the nave of a church from the chancel. It is similar to an iconostasis in an Eastern Orthodox church, except that its origin is more recent. Its construction is different, because it is not a complete visual barrier.

Rood screens are much less common in western churches today than in medieval times, when they originated. Protestants had theological problems with separating the laity from the liturgy. Catholic churches removed the rood screens for the same reason as a result of the Council of Trent (1545–1563). You can still find a rood screen in an Anglican or Lutheran church.

Sacristy   In historic church architecture, the sacristy is the room or closet in which communion equipment, linen, and supplies are kept. It is usually equipped with a sink.

Sanctuary   In historic usage, chancel and sanctuary are synonyms. In Celebratory church architecture, the front part of the church from which the service is conducted, as distinct from the nave, where the congregation sits. The sanctuary is usually an elevated platform, usually three steps up from the nave. In Protestant churches, the term sanctuary is used to mean the place where the congregation sits.

Shrine   A shrine is a building or a place that is dedicated to one particular type of devotion that is limited to commemorating an event or a person. What makes it a shrine is its limited purpose and use. It could be anything from a large building to a plaque mounted on a pole next to the side of the road. A shrine is located on the site where the event occurred that gave rise to the commemoration and the devotion. For example, suppose someone erects a commemorative plaque on the spot where some important person was murdered and people often come there to think about the significance of the event and pray. That is essentially a shrine.

Stage   In Protestant churches where worship can be theatrical and the congregation functions mainly as audience, the architect often enlarges the chancel to accommodate performances and calls it a stage, as in a theater.

Transept   In medieval times, it became necessary to increase space near the chancel to accommodate large numbers of clergy, the choirs, or members of religious orders. The result was a space between the chancel and the nave that extends beyond the side walls, giving the church a cruciform floor plan—meaning that it is cross-shaped when viewed from the air.

Undercroft   Essentially a fancy word for the church basement under the chancel and nave (and transept, if there is one).

PHOTO CREDITS

Cover	**Stephen Titra**
4	Pen and Ink Drawing, Stephen Titra
4.1	Photo, Stephen Titra
4.2	Photo, Stephen Titra
4.3	Photo, Douglas Jones
6.1	Photo, Nathanael D. Jones
7.1	Photo, Stephen Titra
7.2	Photo, Don Washburn
7.3	Photo, Don Washburn
7.4	Photo, Don Washburn
7.5	Photo, Don Washburn
8	Pen and Ink Drawing, Stephen Titra
8.1	Photo, Stephen Titra
8.2	Photo Howard Andersen
11.1	Photo, Douglas Jones
12.3	Photo, Mark Ahlenius
12.4	Photo, Mark Ahlenius
13	Pen and Ink Drawing, Stephen Titra
15.1	Photo, Danley Sound Labs
15.2	Photo, Danley Sound Labs
15.3	Photo, Danley Sound Labs
16	Pen and Ink Drawing, Stephen Titra
16.1	Photo, Nathanael D Jones
18.1	Photo Dale Shirk
18.2	Photo Dale Shirk
18.3	Photo Dale Shirk
18.4	Photo Dale Shirk
18.5	Photo Dale Shirk
18.6	Photo, Douglas Jones

INDEX

Printed and bound by CPI Group (UK) Ltd, Croydon, CR0 4YY

23/10/2024

01778247-0001